Science and Technology

Science and Technology

The Making of the
Air Force Research Laboratory

ROBERT W. DUFFNER

Air University Press
Maxwell Air Force Base, Alabama

November 2000

Library of Congress Cataloging-in-Publication Data

Duffner, Robert W.
    Science and technology : the making of the Air Force Research Laboratory / Robert
W. Duffner.
        p. cm.
    Includes index.
    ISBN 1-58566-085-X
    1. Air Force Research Laboratory (Wright-Patterson Air Force Base, Ohio) I. Title.

UG644.W75 D84 2000
358.4'0072073—dc21

                                                                00-064302

Disclaimer

For Carol, Kelly, Susanne, and Tyler,
Who make life more interesting and worthwhile

Contents

Part 1
The Decision

Part 2
The Transition

Illustrations

Tables

Foreword

History is the study of change. It is an important—but often neglected—resource and tool that allows each of us to analyze and extract the most relevant experiences from the past and apply that knowledge to today's decision-making process. What has happened in the past affects the way we live in the future. Therefore, to ignore history is a mistake. Likewise, to capture the history associated with contemporary events can have huge payoffs for future leaders and is an extremely wise investment of time and energy.

Early in my career, I served on the history faculty at the Air Force Academy, where I challenged cadets to gain a better appreciation for the past. As part of that educational process, I encouraged students to consider history a basic building block in their development as professionals. My goal was to make them more aware of a time-tested database that they, as future leaders, could draw upon in shaping policies and strategy to best accomplish the mission. Later, as chief of staff, I initiated a reading program of selected historical works to promote the professional growth of all Air Force personnel.

This history documents a watershed event within the United States Air Force during my tenure as chief of staff—the creation of the Air Force Research Laboratory (AFRL). As the "high technology" service, the Air Force has always searched for ways to improve continuously its science and technology enterprise. In that context, the making of AFRL was not a bureaucratic accident. Rather, it was the product of a complex mixture of historical forces and pressures at work that convinced people at all levels that the time was ripe to bring about fundamental reform in how the Air Force conducts its business of science and technology.

In terms of significance, a wealth of past studies has focused on almost every aspect of the "operational" side of the Air Force. But there has been a scarcity of available scholarly studies that address the far-reaching implications of science and technology. Bob Duffner's insightful and comprehensive account of the evolution of events leading to the genesis of a single Air Force laboratory is a major contribution that helps fill that gap. Organization and infrastructure are critically important components of the total science and technology picture. Thus, the manner in which our laboratory system is organized is a critical factor in the Air Force's ability to assure that we are investing in and delivering the most relevant technologies possible.

Duffner is an accomplished historian who weaves an engaging and cogent story of how the Air Force moved from 13 separate labs to one consolidated lab. Thoroughly researched and documented, this balanced and highly readable narrative is divided into two parts. Part one addresses the reasons *why* the Air Force decided to consolidate its far-flung science and technology enterprise into one lab. *How* the new lab was implemented is the focus of part two. This study is especially revealing because the reader is given access to the inner workings and struggles of a major Air Force organizational restructuring through interviews with key individuals who participated directly in the decision-making process to establish a single lab.

People—collectively and individually—make history. The creation of the Air Force Research Laboratory represents one of the most sweeping reforms in the history of the Air Force and is testimony to the principle that change is inevitable. Understanding why and how a single lab happened is critically important in assessing where Air Force science and technology has been in the past and where it is going in the future. This book offers a unique perspective on how and why the Air Force altered its organizational approach to science and technology. I strongly recommend that it be added to every serious Air Force professional's reading list.

RONALD R. FOGLEMAN
General, USAF, Retired
Chief of Staff, 1994–97

About the Author

Robert W. Duffner is chief of the Air Force Research Laboratory's history office at the Space Vehicles and Directed Energy directorates (Phillips Research Site) located in Albuquerque, New Mexico. A graduate of Lafayette College, he received his PhD in history from the University of Missouri. He served as an infantry rifle-platoon leader with the 101st Airborne Division in Vietnam and later as a colonel in the Army Reserve while serving as a consulting faculty member at the Army Command and

Robert W. Duffner

General Staff College, Fort Leavenworth, Kansas. Dr. Duffner's earlier book, *Airborne Laser: Bullets of Light* (Plenum Press, 1997), is a history of the world's first successful demonstration of an Air Force laser on board an aircraft that engaged and disabled tactical missiles.

Preface

Vision is often an elusive concept because organizations often package it in terms of an appealing aphorism intended to symbolize efficiency and productivity. But a well-defined vision is useful only if leaders and workers at all levels persistently promote and practice it to bring about fundamental change in how an organization operates. This book is about the creation and implementation of the Air Force's vision to reinvigorate its science and technology infrastructure during the mid-1990s in an effort to keep pace with changing times. This vision, which manifested itself from an organizational perspective in the creation of a single Air Force laboratory, was the latest initiative in an evolutionary chain of activities stemming from the World War II era designed to create, strengthen, and refine the Air Force's research and technology enterprise.

Perhaps no early scientific visionary stood taller than Vannevar Bush, the tough-minded pragmatist who headed the Office of Scientific Research and Development during World War II. Bush's vision was simple but far-reaching. On more than one occasion, he urged President Franklin Roosevelt to ensure the nation's defense by insisting that the government take the lead in funding, promoting, and sustaining scientific research and development *after* the war. In making his point, Bush reminded the president, "If we had been on our toes in war technology 10 years ago, we would probably not have had this damn war."

Gen Hap Arnold and the eminent aerodynamicist Dr. Theodore von Kármán embraced Bush's vision to make science and technology the centerpiece of the nation's airpower strategy. The first step toward implementing this vision involved an in-depth study led by von Kármán that resulted in the publication of *Toward New Horizons* in December 1945. This multivolume report, considered the first comprehensive blueprint for future aerospace development, forecasted those budding technologies that offered the greatest potential for influencing the future of airpower over the next 20 to 30 years. The predominant message was that the United States would have to be willing to take "high risks" and make an unwavering commitment to invest in ongoing research and development programs so the nation would have the most advanced technical weapon systems to fight the next war.

Toward New Horizons and the influence of Arnold and von Kármán helped to spur on a new culture in the military that depended more and more on the contributions of scientists and engineers. This movement toward science and technology gained more momentum in 1947 with the establishment of the Air Force, which almost immediately earned the reputation as the "technically oriented" service. The Air Force's vision of science and technology began to take root quickly with the formation of a permanent Air Force Scientific Advisory Board (1947) and the creation of the Air Research and Development Command (1950), the first command exclusively devoted to advancing science and technology. From this foundation emerged the Air Force Systems Command (1961) and a network of 13 laboratories that remained in existence, first as independent Air Force laboratories and later as a federation of technology centers under which the various laboratories were grouped, until they merged into four major laboratories in 1990. Throughout its first 50 years, the Air Force remained firmly committed to the idea that its research and development infrastructure served as the vehicle for transporting scientific and technological advances to the modern-day battlefront.

By the mid-1990s, the vision of how the Air Force intended to reorganize its science and technology enterprise rested on the shoulders of two men: Gen Henry Viccellio Jr., commander of Air Force Materiel Command (AFMC), and Maj Gen Richard R. Paul, director of Science and Technology at Headquarters AFMC. Working closely together, they initiated a new vision that represented a radical departure from the old way of doing business by establishing a single Air Force Research Laboratory (AFRL) in October 1997. Although the infrastructure changed, AFRL remained true to *Toward New Horizons'* legacy of conducting "high risk" science to produce "revolutionary" technologies.

Six months after the stand-up of the new laboratory, General Paul recognized the importance of capturing the history of the evolution of the single laboratory. I first met General Paul over a scheduled working lunch in his office to discuss what writing a history of this type would involve. After several interruptions by members of his staff, it soon became abundantly clear that every minute of this commander's day was occupied. He reassured me that we would meet again because he considered the history of the lab a worthwhile project. True to his word, we met several weeks later—at a site undisclosed to his staff—during which time I conducted an uninterrupted three-hour interview with him.

That was the start of a one-on-one association which proved invaluable in writing this book. Because of his position as head of Science and Technology at AFMC, General Paul was a central figure in the laboratory story, and I needed to hear his thoughts directly. Fortunately, he was extremely generous in putting aside time to meet and correspond, encouraging me to E-mail or phone him anytime I reached an impasse in the research or writing. I took him up on that offer by frequently pestering him to clarify a variety of minor and major lab issues. I was pleasantly surprised that he thoroughly and completely answered all my E-mails, usually within a few hours and never longer than a day. If he didn't know the answer, he told me he would talk to others to get the information and get back to me—and he always did so. In addition, I greatly appreciated his positive attitude and constant encouragement to move forward with this project. In short, without General Paul's interest and support, I could not have completed this book.

Like General Paul, numerous other AFRL employees gave freely of their time, consenting to interviews and providing information on the evolution of AFRL. Especially willing to help were Col Dennis Markisello, vice commander of AFRL, and Capt Chuck Helwig of AFRL's command section, who furnished over 40 notebooks containing an extensive collection of primary source documents covering various aspects of the lab's development. Dr. Don Daniel, executive director of AFRL, offered an insightful "top-down" look at the lab-reorganization process. Through several interviews, Mr. Tim Dues, with AFRL/Plans and Programs, patiently explained all the inner workings of a complicated laboratory operation. Dr. Brendan

Godfrey, Col Mike Pepin, and Lt Pat Nutz, also in Plans and Programs, furnished useful information on the lab reorganization. Without the assistance of Ms. Bridgett Parsons, AFRL/Human Resources, I never would have been able to locate all the pertinent personnel charts and briefings that addressed the ever-changing personnel picture. Bridgett proved extremely helpful in interpreting a maze of personnel numbers, statistics, and trends. Dr. Hendrick Ruck, AFRL/Human Effectiveness, also supplied excellent input covering the manpower-downsizing plan as the four labs merged into one.

All the AFRL tech directors I consulted were very forthright in providing candid comments on the lab reorganization. Many offered information not readily found in documents: Ms. Christine Anderson (Space Vehicles), Dr. Earl Good (Directed Energy), Dr. Joseph Janni (Air Force Office of Scientific Research), Dr. Vince Russo (Materials and Manufacturing), Mr. Terry Neighbor (Air Vehicles), Dr. Tom Curran and Col John Rogacki (Propulsion), Mr. Les McFawn (Sensors), and Mr. Jim Brinkley (Human Effectiveness). Dr. Robert Barthelemy, who headed the tech directorate team, imparted a wealth of knowledge. Also, Col Mike Heil and Dr. Keith Richey provided unique perspectives on Phillips Lab and Wright Lab, respectively.

Dr. Russo, who directed the lab-transition team, and Ms. Wendy Campbell, his deputy, detailed all aspects of the various task groups that implemented the lab-reorganization plan. They were particularly helpful in explaining the role of the three independent review teams. Dr. Harro Ackermann from Phillips Lab offered sensible information on the day-to-day workings of the lab-transition staff.

Outside AFRL, General Viccellio spoke openly about his motives for reorganizing the laboratory system. His vice commander, Lt Gen Lawrence P. Farrell Jr., provided information about the presentation of the single-lab concept to the Corona meeting in the fall of 1996. As the Air Force's highest-ranking civilian, Dr. Sheila Widnall, secretary of the Air Force, gave her perspective on the single lab. Also, Mr. Blaise Durante, who briefed Secretary Widnall on the final single-lab proposal, offered important insights on how that process worked. His assistant, Lt Col Walt Fred, took the time to locate briefing charts on the series of events that led to the secretary's approval of the new lab. Dr. Gene McCall, who headed the *New World Vistas* study, helpfully explained how outsiders viewed the future of laboratories. Dr. Edwin Dorn, former undersecretary of defense for Personnel and Readiness, furnished his account of the "Dorn cuts" and the way they affected the overall manpower-downsizing plan.

I owe a special debt of gratitude to those steady and consistently productive workers I encounter every day in the history office at the Phillips Research Site at Kirtland AFB, New Mexico. Two reservists were very helpful. Maj Laurel Burnett carefully proofread the entire manuscript and finalized the chronology. Maj Rhonda Toba did an excellent job of organizing and abstracting over 40 lab-management reports stretching back to the 1960s. Ms. Sylvia Pierce put together a detailed time-line chart to illustrate the series of events leading up to the single lab. Our archivist, Mr. Steve Watson, relentlessly contacted a number of government agencies to locate and collect an assortment of critical documents on the single-lab reorganization that were absolutely essential to the narrative. Dr. Barron Oder offered his ideas on content issues and, as our resident computer expert, smoothed out all the pesky computer glitches to ensure that all photos, charts, and so forth appeared in the proper place throughout the text. I also thank Ms. Jessica Gomez, our highly competent stay-in-school employee, who diligently checked the accuracy of all the endnotes.

Finally, Air University Press has been a very supportive partner in this venture. I especially want to thank Dr. Marvin Bassett for his meticulous attention to detail in reading and editing the manuscript. My only gripe with Marvin is that he truly believes the Atlanta Braves are better than the New York Yankees!

ROBERT W. DUFFNER
Kirtland AFB, New Mexico
July 2000

Part 1
The Decision

Chapter 1

Introduction

The first essential of airpower is preeminence in research.

—Gen Henry H. "Hap" Arnold

It was a perfect night for baseball on 9 October 1996. The power-laden Baltimore Orioles had come to do battle with the hometown-favorite New York Yankees in game one of the American League Championship Series. Although playing in the unfriendly confines of the "Bronx Bombers'" ballpark in front of thousands of loyal and screaming New York fans, the Orioles found themselves leading four to three as the Yanks came to bat in the bottom of the eighth. But a dramatic and controversial change of events would wipe out the fragile one-run margin the Orioles had managed to cling to with only one inning left to play.

The rowdy New York crowd was getting restless and louder as Derek Jeter, the rookie Yankee shortstop, stepped to the plate and drilled the first pitch to deep right field—it looked to be a long out. Baltimore right fielder Tony Tarasco backed up to occupy the last two feet of territory on the warning track in front of the wall and reached up to ease the routine fly ball securely into his glove. As it turned out, it was anything but routine. At that same instant, Jeff Maier, an avid 12-year-old Yankee fan, leaned over the outfield wall with his right arm stretched to the limit and managed to "snatch" the ball just as it was about to land in the right fielder's glove. Instinctively, within the blink of an eye, an incredulous Maier jerked his glove and prized souvenir back into the stands. The frenzied stadium crowd went wild with delirium as the umpire signaled "home run" and a bewildered Jeter circled the bases.

The outraged Orioles screamed interference, but the umpire's call stood, and the boy from Old Tappan, New Jersey, suddenly became an instant celebrity in the New York metropolitan area. Jeff Maier's heroic action tied the score and provided just the lift to inspire Bernie Williams to hit the game-winning homer in the bottom of the 11th for a dramatic

3

come-from-behind five-to-four victory for the Yankees in game one. The stunned and tormented Baltimore team never recovered from the unlikely episode of that first game and went on to lose the play-offs, while the Yankees advanced to meet the Atlanta Braves in the World Series.

At the same time the Yankees were in the midst of determining their baseball destiny with the Baltimore Orioles, halfway across the country in Colorado Springs, Air Force leaders were engaged in a series of top-level meetings to map out the nation's aerospace future. There were no Derek Jeters or Jeff Maiers to make the dramatic play in Colorado. Nor was there anything comparable to Bernie Williams's extra-inning heroics at the five-day Air Force Corona conference taking place at the Air Force Academy during the second week of October 1996. Instead, the outcome of the Air Force's game plan and future depended primarily on the decisions made by the Corona attendees. Those seasoned players included Dr. Sheila E. Widnall, secretary of the Air Force; Gen Ronald R. Fogleman, chief of staff of the Air Force; and a select group of other four-star generals who commanded the nine major commands throughout the Air Force. Unlike the Yankees team that could see the fruits of its baseball labors almost immediately, the Air Force squad did not have the luxury of instantaneous feedback and reinforcement. Decisions made by the heady Air Force lineup at Corona '96 focused on long-term global issues affecting the nation's defense that would not be realized for several years down the road.

Although Corona addressed a wide range of topics, one of the most important issues had to do with charting the future course of research and development (R&D) activities within the Air Force. This was of utmost concern to two men: Gen Henry Viccellio Jr., who on 30 June 1995 had assumed command of Air Force Materiel Command (AFMC), headquartered at Wright-Patterson Air Force Base (AFB) in Dayton, Ohio, and Maj Gen Richard R. Paul, who served as the director of science and technology (S&T) under General Viccellio. Although General Paul did not attend Corona '96, General Viccellio did attend and was a major participant because of his position as commander of AFMC. Both men had worked extremely hard for five months prior to Corona, developing and fine-tuning a radically new plan for conducting S&T business. Faced with shrinking budgets and the need to eliminate duplication of effort and similar technological work among multiple labs at different locations, they proposed to consolidate four existing laboratories into one Air Force Research Laboratory (AFRL) designed to lead to a more efficient and streamlined operation.

Viccellio and Paul anxiously awaited the secretary's and chief of staff's reaction to the "single laboratory" proposal. Acceptance of this new plan would have a profound effect on the S&T acquisition process and would mean a complete dismantling of four laboratories to make one. After a short briefing with minimal discussion on the single-lab proposal at Corona, Secretary Widnall and General Fogleman gave their endorsement of the single-lab concept. Final approval would come later, after General Paul prepared a more detailed follow-on briefing scheduled for presentation to Secretary Widnall in her office in November 1996.

Upon hearing of the Corona decision, General Paul realized that this was a tremendously significant turning point for the Air Force's S&T community. But he also realized that this landmark decision was the result of much soul-searching and hard work by him and others to reform the laboratory system by moving off in a totally new direction. He welcomed what he judged to be good news from the Corona meeting, but he was also very much aware that the idea for a single laboratory did not have its origins at Corona. The roots of what would become the Air Force Research Laboratory stretched back to a series of events, requirements, and opportunities that occurred several years prior to Corona's "Gathering of Eagles."

Chapter 2

Rumblings of Laboratory Consolidation

The thought of consolidating laboratories was not new. Over the last decade, this idea had grown out of the Packard Commission's blue-ribbon study (begun in 1985) that looked at ways to operate the Department of Defense (DOD) in a more efficient and economical manner. David Packard, a former undersecretary of defense, headed a high-level team of investigators that focused on four core areas that were candidates for change: national security planning and budgeting, military organization and command, acquisition organization and procedures, and government-industry accountability. Packard's final report, *A Quest for Excellence* (released in June 1986), proposed sweeping reforms, including substantial personnel reductions, to improve efficiency and save money in DOD. President Ronald Reagan directed implementation of the Packard Commission's recommendations in National Security Decision Directive (NSDD) 219, issued on 1 April 1986. The model acquisition-reform plan called for the establishment of "strong centralized policies through highly decentralized management structures."[1]

The 1980s: A Move for Change

The Goldwater-Nichols Department of Defense Reorganization Act, also signed into law by President Reagan in 1986 and considered the most significant defense-reform effort since 1947, was enacted to carry out the reforms proposed in the Packard Commission's report. First and foremost was the creation of a new undersecretary of defense for acquisition (implemented by NSDD 219), whose job was to set overall procurement and R&D policy for DOD. This "acquisition czar" (comparable to the chief executive officer of a major corporation) was the one point of contact who exercised centralized control over all DOD acquisition programs. He exerted influence by establishing a three-tiered program-management system made up of three new service acquisition executives (SAE), one for each military service. Each SAE appointed product executive officers (PEO) to manage a select number of major acquisition programs on a full-time basis. In turn, the PEOs depended on program managers to work individual programs. All this served to establish a more streamlined management system that would trim overhead, eliminate waste, and provide for improved efficiency in directing and monitoring DOD programs.[2]

Initially, the Air Force laboratories were not inclined to show much enthusiasm for supporting the Packard Commission's reforms and the new acquisition-management system enacted by the Goldwater-Nichols Act. Simply put, the labs were reluctant to relinquish control over their R&D programs to the SAEs. By 1989 President George Bush was hearing complaints from congressmen who were not pleased that the services seemed to be dragging their feet in getting behind the management reforms initiated by the Packard Commission and Goldwater-Nichols Act. In February 1989, in response to congressional pressure and budget reductions proposed for defense in the Gramm-Rudman-Hollings Act, President Bush directed Secretary of Defense Dick Cheney to draft a plan to look at ways to improve management (with fewer employees) and organizational efficiency in DOD. One of Cheney's major challenges entailed devising a strategy to fully implement the sweeping DOD reforms proposed in the Packard Commission's report.[3]

On 12 June 1989, Cheney completed a major reorganization plan known as the Defense Management Review (DMR), which addressed ways to improve the defense procurement process and urged the military services to borrow and implement streamlined business practices used in the private sector. The benefits of applying the most acceptable, time-tested business methods to defense activities included more effective and efficient operations, reduced costs, and a higher level of job satisfaction by military and civilian employees, whose ratings would depend upon performance rather than longevity. Cheney reported to the president that, with appropriate congressional legislation, the DMR would be able to fully implement the Packard Commission's recommendations. Perhaps this thinking was optimistic, but it represented a major commitment by the secretary, signaling that he and the Bush administration were serious about implementing substantial changes in the way DOD conducted its daily business.[4]

Why did DMR recommend such drastic changes? Secretary of the Air Force Donald B. Rice believed that the answers were simple—DOD could not continue to conduct "business as usual" because budgets would clearly get smaller in the out years. Congress and taxpayers demanded this change as the cold war wound down and external threats appeared to diminish. But at the same time lower budgets were becoming a fact of life, Congress knew it still had to produce the maximum amount of military capacity possible to deal with any impending regional crisis. One way to meet declining future budgets was to begin to reduce the number of military personnel and civilians working for DOD. More specifically, Rice and others in the highest leadership positions took an aggressive stance to mount an "attack" on overhead. According to Rice, "It's simply imperative that we resize support and overhead to suit the reduced military capability rather than continue to do business as usual."[5]

Overhead appeared in many shapes and forms. Some jobs, simply less essential than others in both the support and technical areas, could be eliminated to help reduce costs. Reducing bureaucratic layering by doing away with two headquarters could also produce savings (e.g., DMR's recommendation to consolidate Air Force Systems Command with Air Force Logistics Command). Although the Air Force at first strongly resisted this type of action, such a merger would result in lowering overall personnel numbers as well as moving toward a more efficient organization by pushing more responsibilities down the chain of command to subordinate units. DMR's recommendation to consolidate the two commands was not idle talk. In anticipation of a possible merger, Secretary Rice remarked that "it's always better to do it yourself" rather than have congressional action force the decision. To underscore the idea that fundamental changes had to be made to get away from doing business as usual, the Air Force took the initiative and made a bold move to implement DMR thinking in a relatively short time. On 1 July 1992, Systems and Logistics Commands merged to form the newly activated Air Force Materiel Command (AFMC).[6]

As part of the continual defense-management review process that followed his DMR report of June 1989, Cheney appointed groups to investigate ways of consolidating defense functions, including laboratories. Defense Management Report Decision (DMRD) 922, issued on 30 October 1989, included one of the findings of these special study groups. This decision strongly advised that the Pentagon give serious consideration to merging all military labs directly under DOD. Some people envisioned that a single DOD laboratory was not out of the question since such a lab would become the institutional focal point for all of DOD's future S&T matters. Advocates also reasoned that establishing a single laboratory was the best way to proceed because too much overlap and duplication existed among the service labs; consolidation would cut overhead and operating costs, strengthen DOD's technology base with fewer and larger laboratories, and boost productivity and efficiency. Other possibilities included intraservice and interservice consolidations, providing for "lead laboratories," headed by the Army, Navy, or Air Force, that would focus on technology common to all three military services.[7]

DMRD 922 directed John A. Betti, undersecretary of defense for acquisition, to conduct an extensive study that focused on the advantages and disadvantages of interservice and intraservice consolidation of laboratories. Betti called on Dr. George P. Millburn, deputy director of defense research and engineering in the Pentagon, to work with the three services to explore the entire range of laboratory options. On 30 April 1990, Millburn reported his findings to Betti, recommending a number of possible solutions, including reducing the number of labs and combining all service labs into one DOD laboratory. Betti still did not have a definitive answer on how to proceed in the future structuring of labs, preferring to postpone his decision until he had more facts. Consequently, he instructed each service to come up with its best recommendation on what to do with the labs.[8]

The Air Force and Systems Command knew they had to come up with a realistic plan of action or be at the mercy of DOD. If they did not present their own internal plan, then Betti would likely make the decision for them and support the idea of combining all service labs into a single, centralized DOD lab. That was too much of a gamble for the Air Force to take. Because Systems Command did not want to get caught in a winner-take-all game, the Air Force decided to compromise. By proposing a consolidation of its 13 labs (as well as the Rome Air Development Center) into four labs, Systems Command hoped it could duck DOD's single-lab bullet—at least temporarily.

Hopefully, the creation of four superlaboratories would silence proponents of the single DOD lab for the short term. Critics argued that 13 separate labs were too small to perform "world class" science because they lacked the synergy and interdisciplinary expertise that made America's large national laboratories so successful. These model labs—Sandia, Los Alamos, and Lawrence Livermore—had established enviable international reputations over the years because of their consistently high level of performance. Further, some DOD leaders thought that the Air Force maintained too many labs that were too geographically dispersed; this kind of poor management resulted in a wasteful duplication of technical efforts that cost the government more money than necessary. Systems Command, however, hoped that its proposed consolidation would satisfy the critics by establishing a more centralized span of control that would weed out unproductive technical efforts.

Also, fewer labs meant fewer people, which translated to additional savings. For years the military services were notorious for representing most staff functions at each level of the organizational structure. Thus, the military resembled a giant octopus with its staff-function tentacles reaching out and strangling every organizational level, regardless of the size of the operation. Private business, driven almost exclusively by profits, considered such excessive support detrimental to cost-effectiveness. DOD and the Air Force, both committed to mirroring the organizational techniques of successful businesses, realized they had to mount and sustain an assault on overhead if they expected to effect any meaningful change. Theoretically, four labs would require only four support staffs instead of 13 and, obviously, fewer people. The standing up of four Air Force laboratories on 13 December 1990, along with the combining of Logistics and Systems Commands in July 1992, attested to the Air Force's resolve to take steps to show DOD leaders that it was serious about reducing personnel numbers and eliminating the duplication of technology efforts (table 1).[9]

Table 1

Consolidation of 13 Air Force Laboratories

Air Force Space Technology Center
Kirtland AFB, New Mexico

  1. Weapons Lab, Kirtland AFB
  2. Geophysics Lab, Hanscom AFB,   →   Phillips Laboratory,
     Massachusetts   Kirtland AFB
  3. Astronautics Lab, Edwards AFB,
     California

Wright Research and Development Center
Wright-Patterson AFB, Ohio

  4. Avionics Lab, Wright-Patterson AFB
  5. Electronics Technology Lab,   →   Wright Laboratory,
     Wright-Patterson AFB   Wright-Patterson AFB
  6. Flight Dynamics Lab,
     Wright-Patterson AFB
  7. Material Lab, Wright-Patterson AFB
  8. Aero Propulsion and Power Lab,
     Wright-Patterson AFB
  9. Air Force Armament Lab,
     Eglin AFB, Florida

Rome Air Development Center   →   Rome Laboratory,
Griffiss AFB, New York   Griffiss AFB

Human Systems Division
Brooks AFB, Texas

10. Air Force Human Resources Lab,
     Brooks AFB
11. Harry G. Armstrong Aerospace   →   Armstrong Laboratory,
     Medical Research Lab,   Brooks AFB
     Wright-Patterson AFB
12. Air Force Drug Testing Lab,
     Brooks AFB
13. Air Force Occupational and
     Environmental Health Lab,
     Brooks AFB

Although this consolidation process satisfied the Air Force's short-term interests, the service did not fully realize the long-term consequences of this dramatic event. Air Force leaders had unwittingly planted seeds that would grow into a single Air Force laboratory in 1997. Many people perceived the laboratory reform movement of the mid-1990s as a logical extension of the DMR process that had led to the formation of the Air Force's four superlabs in 1990. Beyond that, events generated by the Clinton administration and Congress would become constant reminders that set the mood and kept the pressure on in the 1990s for moving one step closer to what many top-ranking officials predicted would be the formation of a single DOD laboratory after the turn of the century.

President Clinton and Laboratory Reform

Consistent with his campaign promises, President William Jefferson Clinton pledged to work to ensure that the nation sustained its position of world leadership in science and technology. On 23 November 1993, he established the National Science and Technology Council (NSTC) and announced an ambitious plan to undertake an across-the-board review of all federal laboratories. His basic goal was to streamline laboratory operations in view of the projected decrease in federal R&D dollars. Although he recognized that less money would be available, President Clinton refused to sacrifice the quality of S&T programs. He believed that, by reorganizing and consolidating selected parts of the laboratory system, people would work smarter and S&T would not suffer.[10]

Over the next six months, the executive branch became more and more active in working toward turning the president's ideas into reality by sponsoring a major effort to reevaluate the operation of the government's laboratories. Clinton's interest and desire to have his administration take the lead in examining how the nation conducted its S&T business were only one part of a two-pronged, pressure-driven strategy to reform DOD laboratories. Although no one foresaw such a development, the president's initiative for laboratory reform would merge two years later with Congress's own aggressive

13

plan to revitalize and reduce the number of labs. Congressional language describing this plan was spelled out in the National Defense Authorization Act passed in February 1996, which served to "kick-start" DOD to move out on the laboratory-reform issue. Out of this legislation evolved *Vision 21*, DOD's long-range plan for making fundamental changes to the military labs (see chap. 3 for a more detailed treatment of *Vision 21* and the National Defense Authorization Act).

On 5 May 1994, President Clinton issued a directive establishing the Interagency Federal Laboratory Review. To get this process under way, he instructed NSTC to review the nation's three largest laboratory systems operating within DOD, the Department of Energy, and the National Aeronautics and Space Administration. These three laboratory systems accounted for roughly one-fifth of the federal government's total investment ($15 billion out of $70 billion) in R&D. With the end of the cold war, the president and Congress realized that the roles, missions, and responsibilities of the laboratories had to change rapidly to meet a new set of global threats and requirements.[11]

A reassessment of the continuing value of laboratories in serving vital public interests was a logical extension of the reform-minded National Performance Review (renamed the National Partnership for Reinventing Government) of 1993, headed by Vice President Al Gore and designed to create a smaller and more cost-effective government. Clinton considered this program one of the top priorities of his first administration and made a special effort to publicly support and praise Gore for his determination to eliminate waste and inefficiency throughout the government. As part of this overall commitment, the president wanted to find out what specific "options for change" within the laboratories might be available to cut costs and at the same time improve R&D productivity. In other words, in this era of making government smaller and more efficient, the president and his team strongly urged that all federal agencies "do more with less." Although many people considered this a legitimate goal, others had doubts and wondered in private about the reality of doing more with less. When it came down to day-to-day working conditions, people at the working levels oftentimes sarcastically interpreted this to mean "doing less with less."[12]

The president was quick to point out that, over the years, the laboratories had maintained an extraordinary and highly talented workforce that had established a strong track record of providing the nation essential services in fundamental sciences. Not only had the labs "contributed greatly" in the past, but the nation counted on them to make even more significant contributions of "tremendous importance" in the future. Certainly, Clinton did not advocate shutting down the laboratories. His main objective was to make them run as efficiently as possible. Although he gave no specifics as to what changes would have to be made, he hinted at taking steps to streamline management before sacrificing R&D programs. Most people interpreted this to mean that the number of people working at the various labs would have to be reduced in order to achieve budget savings projected over the next few years.[13]

After a year of investigating how the laboratories operated, on 15 May 1995 NSTC submitted to the president its final report, confirming what most people associated with the labs had already known. DOD labs were specialized institutions that zeroed in on S&T programs designed solely to enhance the war-fighting capability of the nation. What they researched, developed, and transitioned in terms of hardware to the operational fighting forces became one of the most enduring pillars of the country's national security policy. NSTC recognized that the overriding mission of the DOD laboratories was to strengthen national security by advancing technology, which in itself was a compelling argument to retain a quality laboratory system that could effectively serve the changing security needs of the nation. The legitimacy of laboratory functions enjoyed unanimous support, but NSTC also concluded that ample room existed to improve management and cut redundancy throughout the laboratory system.[14]

The NSTC report's fundamental recommendation emphasized that, to achieve greater efficiency in government, DOD had to come up with a realistic laboratory-restructuring plan to implement cross-service integration and maximum use of common support assets. This meant eliminating the duplication of effort by consolidating lab resources. Although the Defense Base Realignment and Closure Commission of 1995 (BRAC 95) had also advocated combining similar R&D programs under one laboratory, only limited progress had occurred in meeting the specific goal of cross-service integration. BRAC did not produce a major laboratory consolidation move, but in 1995 NSTC and the president remained confident that similar lab functions needed to be combined because of basic changes taking place in DOD. First and foremost, the downsizing of DOD labs would definitely occur. Declining budgets and a shift in missions—the result of the changing threat brought about by the end of the cold war—meant less demand for new acquisitions systems for advanced weapons traditionally developed by military labs. In fact, prior to the NSTC report, the president and Congress had already mandated, as part of across-the-board downsizing in government, the phasing in of a 35 percent reduction in laboratory personnel from 1994 through 2001. Cross-service integration of lab functions, the NSTC reasoned, would help accomplish the goals of reducing staff and budgets and at the same time retain a quality lab infrastructure to continue meeting future mission requirements.[15]

After reviewing the NSTC final report, President Clinton issued further direction and guidance to DOD to continue pursuing alternatives that would achieve consolidation of the labs. Specifically, he instructed the secretary of defense to submit a report to him by 15 February 1996 "detailing plans and schedules for downsizing the DoD laboratories. This report was to identify opportunities for greater efficiency through measures such as cross-service integration and service lab consolidations." The president intentionally did not get bogged down in outlining the exact details of specific changes needed in reforming the labs. However, he did present four general guidelines and principles for DOD to follow in preparation of this final report.[16]

One fundamental area requiring immediate attention called for DOD to develop a workable game plan to reduce the excessive amount of paperwork—internal management instructions, regulations, policy procedures, and so forth—that absorbed an enormous amount of time and impeded laboratory performance. A second pressing issue entailed a reexamination and clarification of each laboratory's mission to eliminate re-dundancy and thus help restructure the entire lab system. A third topic worthy of close scrutiny involved reducing or elimi-nating low-priority R&D programs as another way to improve efficiency. Finally, DOD had a responsibility to explore potential opportunities to "coordinate and integrate laboratory resources and facilities on an interagency and inter-service basis, eliminat-ing unnecessary duplication and establishing joint manage-ment where appropriate." In other words, to achieve more cost-effectiveness, would it be better to combine parts or all of the military labs under one consolidated DOD laboratory?[17]

The Setting: Changing the Laboratory Image

No single event or landmark decision was responsible for the establishment of the Air Force Research Laboratory on 31 October 1997. Rather, a combination of events, decisions, and other forces spread over several years led to the creation of the new laboratory. Much of the apprehension, uneasiness, and unknown aspects that attended the process of completely re-structuring the Air Force's four-laboratory system came about because of a continual stream of mandated guidance and pressures imposed on the Air Force by President Clinton, Con-gress, and DOD. By the mid-1990s, the time for planning, studying, and assessing the laboratory structure was clearly over, for the most part. The Clinton administration, Congress, and DOD wanted the Air Force to take action and start recon-figuring the labs to cut more fat and produce an even leaner and more cost-effective R&D operation.[18]

The critical issues driving the move to reorganize the four Air Force laboratories scattered across the country included addressing matters of money and personnel, as well as finding a more efficient way to advance the type of S&T that produced superior products for the war fighter. The image and reputa-tion of the labs had become somewhat tarnished in the 1990s. Many people, both inside and outside the Air Force, perceived each of the four labs as a powerful and independent institu-tion in its own right that relentlessly protected its technology turf. They noted the ever-present and seemingly unresolvable

17

issue of "stovepipes" and "seams" (i.e., duplicating research and technology efforts among several labs at different locations rather than concentrating in one centralized organization at one location). Eliminating these stovepipes and seams proved difficult because each lab lobbied vigorously to defend its portion of the technological pie. Moreover, the four-lab organizational structure did not lend itself to cross-communications among labs—that is, crossing the organizational seams and communicating from one stovepipe to another.[19]

In an effort to formulate a nonduplicative, integrated S&T investment across all four labs, the Air Force created the Air Force technology executive officer (TEO). While each lab reported to a parent product center (i.e., Wright Lab to Aeronautical Systems Center, Phillips Lab to Space and Missile Systems Center, Rome Lab to Electronics Systems Center, and Armstrong Lab to Human Systems Center), the TEO worked directly with the four lab commanders on S&T investment-strategy issues to ensure a cohesive, integrated budget that addressed the needs of a broad customer base. The TEO, in turn, reported to the secretary of the Air Force for acquisition (SAF/AQ), who also served as the Air Force's acquisition executive. Thus, the arrangement somewhat resembled the PEO arrangement mandated by the Goldwater-Nichols Act of 1986, wherein program managers reported to PEOs, who in turn reported to a service acquisition executive.[20]

Under this framework, General Paul was "dual hatted" as Headquarters AFMC's director of S&T (reporting to the AFMC commander) and as the Air Force's TEO (reporting to SAF/AQ). Likewise, each of the four lab commanders had two bosses: (1) the TEO (a two-star) for investment and budget issues and (2) his or her respective product-center commander (a three-star) for manpower, facility, and other infrastructure issues (with the product-center commander writing the lab commander's annual performance rating). The TEO arrangement had definite advantages in promoting an integrated S&T investment strategy, but some people questioned whether it was optimal in light of the four lab commanders' dual-reporting channels and the fact that no single authority had accountability for the full set of S&T resources (money, people, and facilities).[21]

Many people also had the impression that the labs were over-stocked with staff and support people. The tooth-to-tail ratio was out of balance in terms of the proportion of scientists and engineers (the tooth) versus staff and support personnel (the tail) assigned to the labs. For every two scientists assigned, there was one support staff person in place, a ratio that needed adjusting more in favor of the number of scientists. Despite the legitimacy of this concern, in fairness, all the labs since 1990 had gradually reduced the number of people assigned. From 1990 through 1996, total laboratory manpower decreased from 8,480 to 7,226—a 15 percent reduction, largely from the support staff. This gradual drawdown of the laboratory workforce since 1990 occurred because of various mandated personnel reductions levied on the labs each year by DOD and the Air Force. However, one common complaint was that the process cut too much muscle from the labs, since a portion of these reductions involved the elimination of scientist and engineering positions. A second criticism was that, even though each of the four labs' manpower numbers had declined since 1990, DOD officials were convinced that in the long run—based on projected budget reductions for the future—they would have to remove a larger number of positions from the labs' manning documents and at a much quicker pace.[22]

Thus, when Headquarters Air Force and DOD looked at the labs, they saw four fully staffed and fiercely independent organizations determined to retain their separate identities. Over the years, each lab had worked hard to establish a unique and, in many cases, narrow customer base within the Air Force. Further, each lab nurtured this relationship, knowing that it could offer its customers scientific and technical services that no other lab could provide. In essence, because of its specialized expertise, each lab had cornered and monopolized a portion of the R&D market and wanted to make sure that relationship continued to grow and prosper.[23]

Under the organizational structure from 13 December 1990 until 31 October 1997, each lab, to a large degree, remained autonomous. Its future depended on how it developed and marketed its own technology products, with little regard to what the other labs were doing and no incentive for working closely with them. Additionally, the fact that the labs were geographically separated by hundreds of miles made it difficult to work closely together on a day-to-day basis. Each lab focused on building existing programs, attracting new programs, enhancing its own reputation, and conveying to the Air Force leadership that it could transition significant advances in technology to the operational commands. Yet, some high-level officials responsible for managing the laboratory system, especially in this new era of downsizing, began to question and reevaluate what some people perceived to be the inflated contributions of the laboratories across the board. Decision makers, who now began to step back and rethink the laboratories' role, perhaps could take to heart the advice Theodore Roosevelt offered nearly a century ago: "I did not care a rap for the mere form and show of power; I cared immensely for the use that could be made of the substance."[24] Substance was what the Air Force leadership was looking for as it began taking steps to reshape its four separate laboratories into a more coherent, focused, and synergic organization.

Notes

1. President's Blue Ribbon Commission on Defense Management, *A Quest for Excellence: Final Report to the President* (Washington, D.C.: President's Blue Ribbon Commission on Defense Management, 30 June 1986), xii, 55; and Lawrence R. Benson, *Acquisition Management in the United States Air Force and its Predecessors* (Washington, D.C.: Air Force History and Museums Program, 1997), 40–41.

2. *Goldwater-Nichols Department of Defense Reorganization Act of 1986,* Public Law 99-433, 99th Cong., 2d sess., 1 October 1986; John T. Correll, "Streamlining with a Splash," *Air Force Magazine,* March 1990, 12–13; Office of the Secretary of the Air Force, "Seamless Life-Cycle Process: A Vision for Air Force Acquisition in 2050," draft, 14 June 1996; and Benson, 41–42. The new acquisition plan called for military acquisition staffs at service headquarters to be eliminated and consolidated as part of the service secretariats.

3. Dick Cheney, *Defense Management: Report to the President* (Washington, D.C.: Department of Defense, July 1989); and *The Balanced Budget and Emergency Deficit Control Act of 1985,* Public Law 100-119, 100th Cong., 1st sess., 29 September 1987. This act, better known as the Gramm-Rudman-Hollings Act, provided for automatic spending cuts if Congress failed to reach established budget targets. After automatic cuts were declared unconstitutional, a revision of the act in 1990 changed the focus from deficit reduction to spending control.

4. Cheney; and *The Balanced Budget and Emergency Deficit Control Act of 1985.*

5. Donald B. Rice, memorandum to Maj Gen C. P. Skipton, subject: Defense Management Review, 21 August 1989.

6. Ibid.; Larry Grossman, "Re-Winging the Air Force: Streamlining for Leaner Times," *Government Executive Magazine,* December 1991, 11–14; and Benson, 42–50.

7. Briefing, Maj Gen Robert R. Rankine Jr., Air Force Systems Command (AFSC)/XT, subject: Horizons South 90: Science and Technology, 7 June 1990; Gen Ronald W. Yates, AFSC/CC to Aeronautical Systems Division (ASD)/CC et al., letter, subject: Laboratory Consolidation, 1 June 1990; and message, 142130Z DEC 89, AFSC/ST to Air Force Office of Scientific Research (AFOSR)/CC et al., subject: OSD [Office of the Secretary of Defense] Laboratory Consolidation Study, December 1989.

8. Rankine; Yates; and AFST/ST message.

9. Rice.

10. Executive Order 12881, Establishment of the National Science and Technology Council, 23 November 1993; and White House Press Release, Office of Media Affairs, "Statement of the President" [National Science and Technology Council], 23 November 1993.

11. White House Press Release, Office of the Press Secretary, "Statement by the President: Future of Major Federal Laboratories," 25 September 1995. The official name of the president's directive was Presidential Review Directive/National Science and Technology Council (PRD-NSTC-1).

12. Ibid.

13. Ibid.

14. Report, "Future of Major Federal Laboratories: Findings," The White House, Office of the Press Secretary, 25 September 1995; on-line, Internet, 28 April 1996, available from http://www.dtic.mil/labman/fedlabs/toc.html.

15. Report, "Future of Major Federal Laboratories: Recommendations," The White House, Office of the Press Secretary, 25 September 1995; on-line, Internet, 28 April 1996, available from http://www.dtic.mil/labman/fedlabs/toc.html.

16. White House Fact Sheet, "Federal Laboratory Reform," 25 September 1995.

17. Ibid.

18. Maj Gen Richard R. Paul, commander, Air Force Research Laboratory, interviewed by author, 2 March 1998; and Dr. Don Daniel, executive director, Air Force Research Laboratory, interviewed by author, 27 July 1998.

19. Paul interview.

20. Notes, Maj Gen Richard R. Paul, 7 January 2000.

21. Ibid.

22. Ibid.; and fax with attached briefing chart, "Laboratory Manpower (FY 89–01)," Bridgett Parsons, AFRL/HR, to author, 7 April 1998.

23. Tim Dues, associate director, Plans and Programs Directorate, Air Force Research Laboratory, interviewed by author, 2 and 12 March 1998 and 6 April 1998.

24. Theodore Roosevelt, *Selected Works of Theodore Roosevelt,* vol. 5, *Theodore Roosevelt: An Autobiography* (New York: Macmillan, 1913), 357.

Chapter 3

The Catalyst: National Defense
Authorization Act and *Vision 21*

One of the most demanding and persistent challenges with
which General Paul and others in the Air Force R&D community continually had to contend was the steady pressure
brought on by numerous studies recommending that the laboratories reorganize. For years, senior leaders in the Air Force
and DOD recognized that these studies collectively advocated
and hammered home the point that laboratories had to reduce
personnel and budgets, as well as eliminate waste and duplication of technical efforts, if they expected to survive into the
twenty-first century. In essence, the upper echelons of government said that the Air Force would have to make dramatic
changes in the way its laboratories operated.[1]

The problem was that most of these studies offered conclusions and recommendations that were often broad in scope or
focused on only a few specific aspects of laboratory operations.
No single, all-encompassing blueprint outlining the details of
how the laboratories should reorganize existed. However, everyone generally agreed that the labs would have to discard some
of their personnel overhead and at the same time become more
proficient in implementing better business practices for conducting and managing their S&T programs. In other words,
change was inevitable.[2]

General Paul knew better than anyone else that something
had to be done soon to revive the labs. By the mid-1990s, the
time had come to make some hard decisions and move forward. For several years, Paul had wrestled with the problem,
assessing numerous schemes for restructuring the laboratory
system. Part of his preliminary mental gymnastics involved
conceiving different consolidation strategies that included going from four labs to three, to two, and even to one lab. Although he deliberated long and hard about coming up with
better ways for restructuring the laboratory system, he basically kept these ideas to himself during the early stages of his
thinking. He knew that something would have to be done in

Maj Gen Richard R. Paul proposed the single-lab concept.

the long run to respond to the nagging message that laboratories would have to change. But no compelling or immediate urgency to do this presented itself just yet. In his mind, some reform was already taking place since each lab had annually contributed its fair share of personnel reductions, the main portion of which became known as the "Dorn cuts" (after Edwin Dorn, undersecretary of defense for personnel and readiness, who had levied these reductions on the Air Force labs in June 1994). However, many people believed that the Dorn cuts were only a Band-Aid solution to a much deeper laboratory problem.[3]

By February 1996, circumstances had changed dramatically to force Generals Paul and Viccellio to revisit the laboratory options and make some tough decisions. The single most important event was passage of the National Defense Authorization Act for Fiscal Year 1996 on 10 February 1996. Section 277 of this legislation (Public Law 104-106) contained unequivocal language directing the secretary of defense to develop a five-year plan to consolidate and restructure laboratories and test and evaluation (T&E) centers assigned to DOD. Congress directed that "the plan set forth the specific actions needed to consolidate the laboratories and test and evaluation centers into as few laboratories and centers as is practical and possible, in the judgment of the Secretary, by 1 October 2005."[4]

To get this process under way, the National Defense Authorization Act instructed the secretary of defense to submit an initial plan outlining strategy for accomplishing the consolidation and restructuring of the labs and test centers to the congressional defense committees for their review no later than 1 May 1996. Delivered to Congress on 30 April 1996, this first plan—*Vision 21*—became the catalyst for the Air Force's complete revamping of its laboratory system.[5]

As a first step to develop a timely response to the laboratory portion (as opposed to T&E centers) of section 277, Dr. Lance Davis, who headed the Laboratory Management Section of DOD's Office of the Director, Defense Research and Engineering, convened a working group to collect information to clarify congressional expectations. General Paul served as a member of this group, which made a visit on 6 February 1996 to congressional staffers responsible for the language that appeared in section 277. The purpose of the visit was to get the staffers' candid interpretation of the legislation in terms of plans for the laboratories. Led by Dr. Davis, the group met separately with Bill Andehazy, senior staffer on the House National Security Committee, and John Etherton, senior staffer on the Senate Armed Services Committee.[6]

Much of the information that came out of the meetings with the two staffers focused on rather lofty goals. For instance, they believed that too many facilities (labs) remained open and that laboratory infrastructure (defined as bricks and mortar) could be reduced by 50 percent. This number seemed unrealistic to members of the working group, especially since DOD's BRAC exercise in 1995 had had only modest success in closing bases. Although both staffers said their intent was not to "purple-ize the labs" by putting them all under direct control of DOD, they did believe it would be possible to achieve physical consolidation of two geographically diverse facilities. Reminded that BRAC 95 did not seem to work in terms of consolidating activities and organizations, the staffers bluntly replied, "Go back and try again."[7]

General Paul walked away from these meetings convinced that radical changes would have to take place within the Air Force labs in order to meet the expectations of Congress. After returning to Wright-Patterson AFB, he began weighing all the possible implications of the National Defense Authorization Act, revisiting in his own mind the assets and liabilities of creating a single Air Force laboratory.[8]

With the emergence of the National Defense Authorization Act, the spring of 1996 was a tense time, requiring cool heads for making decisions of substantial consequence on the future of laboratories. If changes were to come, then General Paul, who had spent almost his entire career working in the Air

Force R&D area from the ground up, was the right man at the right time to lead this conversion process. He understood how the system worked by virtue of his assignments at two Air Force laboratories (Weapons Lab and Wright Lab), a product center (Electronic Systems Division), two command headquarters (Strategic Air Command and Air Force Materiel Command), and Headquarters Air Force, as well as a Joint Staff assignment. Not only did he have an enviable operational record that helped him build his lab savviness, but also he had impressive academic credentials, having graduated from the University of Missouri at Rolla with a degree in electrical engineering and from the Air Force Institute of Technology with a master's in electrical engineering. Taking his professional military education studies seriously, he excelled in the classroom and was named a distinguished graduate of Squadron Officer School, Air Command and Staff College, and the Naval War College—a feat few officers could match.[9]

His two most recent assignments at Wright-Patterson gave him an opportunity to observe and participate in the decision-making process at the top of the S&T pyramid in the Air Force. As commander of Wright Laboratory from July 1988 to July 1992, he gained invaluable, practical, day-to-day experience in directing major technology programs that advanced aerospace systems. Selected to serve as director of S&T at Air Force Materiel Command in July 1992, he was responsible for leading and devising investment strategy covering the full spectrum of Air Force technology activities. These last two assignments, more than any others, allowed him to see firsthand the fundamental problems of the laboratories. Above all, this specialized on-the-job experience gave him some lasting insights that he used to formulate basic, commonsense principles that he would apply to improve the day-to-day operations of laboratories.[10]

But it took more than formal education and the right on-the-job experience to make a difference in convincing others to move the laboratory system in an entirely new direction. Personality entered the equation as well. As with any respected leader, the one personal quality that stood out above all others was General Paul's tenacious work ethic. People who worked with him and knew him well were amazed by his stamina. He often set a frantic pace to complete complicated taskings but never demanded more from those who worked for him than he was willing to shoulder himself. As many of his staff members pointed out, General Paul remained intensely focused and worked long hours. Very detail oriented, he left his office at the end of the day with a briefcase—sometimes two—full of papers to review well into the late evening hours to ensure he was fully prepared to deal with all the anticipated exigencies of the next day. As one associate put it, "You just couldn't give him a glossy overview briefing—he wanted the specific facts and details to later help him make an informed decision."[11]

Dr. Donald C. Daniel, General Paul's right-hand man

The general did not shy away from making decisions. By nature, he was more analytical than emotional in surveying problems and coming up with solutions. Neither impulsive nor the type to jump to conclusions quickly, he invested an enormous amount of time, effort, and energy collecting and sifting through all possible options—perhaps too much time in the eyes of some people. Moreover, he was most comfortable and confident developing "incremental steps" in breaking down and assessing all the available data when trying to devise and settle on final solutions. Generally, he liked to sort things out in his own mind before consulting with others. Although he was not averse to working alone, when he needed advice, he turned first to his closest and most trusted advisor, Dr. Don Daniel. Under General Paul, Daniel served as both deputy director of S&T and as AFMC's chief scientist.[12]

A first-rate scientist who over the years had earned the reputation of being a no-nonsense and extremely effective manager of R&D programs, Daniel was hard nosed, highly disciplined, and result oriented. He was driven to excel, proud of his accomplishments, and not a man to cross. Daniel honed his scientific and management skills through 16 years of work at the Air Force Armament Center, leading a number of programs, mainly in the areas of aeromechanics and aerodynamics. In 1988 he moved up to become chief scientist of the Arnold Engineering Development Center. Four years later, he joined General Paul's team. Clearly, Paul ran the organization, but Daniel's value was that he never hesitated to present his candid and often opposing opinions to General Paul. Without a doubt, Daniel was Paul's "right-hand man" whom the general used as a sounding board and an honest broker to provide his perspective on major and minor decisions. Paul valued Daniel's judgment and wealth of experience; together they formed a formidable team.[13]

When all was said and done, Paul made the final decision. Once he formulated his new laboratory plan and made up his mind to implement it, there was no turning back. This revealed the dual side of his personality. On the one hand, he was the visionary who thoroughly thought out and devised the big game plan, but his involvement did not end there. Equally important to him was the implementation and successful completion of his vision. In effect, he operated as both the coach and quarterback to accomplish the goals he set. In the case of moving to a single lab, once the decision was made, Paul became a remarkably active participant in making sure the new laboratory stayed on schedule. To Paul, vision and a well-thought-out game plan were the essential first steps to bring about fundamental changes. But he also realized that the best vision and game plan meant nothing unless the team could push the ball across the goal line.

Although making changes of this magnitude proved stressful, Paul worked smartly to be sensitive to the needs of a diversity of people at all levels. Leadership by intimidation was not his style. Instead, he relied on logic and persuasion to get things done. He never abused the power of his rank or office—his two stars did not interfere with his ability and desire to connect with people at all levels to gain additional perspective and knowledge. A man who confronted enormous job pressures with remarkable composure, Paul understood the virtue of patience. In spite of an unrelenting schedule, he made every effort to listen and respond directly and honestly to the concerns of individuals and special-interest groups at all levels throughout the organization. His optimism, unforced smile, and friendly personality contributed immensely to putting people at ease; thus, in the long run he gained their support and confidence as the laboratory marched into a new era fraught with uncertainty.

General Paul recalled that the first step in motivating him to think seriously about proposing a single lab came about because of the congressional language in section 277 of the National Defense Authorization Act, passed in February 1996. The second major event that substantially influenced his decision to form one laboratory was the submission of *Vision 21* by the secretary of defense to Congress on 30 April 1996. Paul used the critical three months between these two very closely related events to clarify in his mind that proposing to consolidate all laboratory assets in one organization was the most logical way to proceed. Paul's plan staked out only the Air Force's internal strategy as a way to respond to reducing infrastructure costs at the labs. An important part of this radical plan was that Paul and others realized from the start that infrastructure included not only real estate and facilities but also people.[14]

The genesis of Paul's proposal for a single lab took place on an airplane on which he and General Viccellio were returning from a temporary-duty trip in the late spring of 1996. At this time, Paul first approached Viccellio about a single laboratory. In light of the National Defense Authorization Act of 1996 and *Vision 21,* Viccellio, as AFMC commander, realized he eventually had to come up with a substantive game plan to address the laboratory problem. During this one-on-one discussion that began to explore potential options, Paul explained the many advantages of going to a single lab. Concentrating all Air Force lab resources under one organization made sense because this approach would streamline R&D programs, make it easier to reduce overhead personnel, eliminate fragmentation of similar technologies currently distributed among multiple technology directorates at various locations, and become more cost-effective. In addition, they had to consider political ramifications. Forming one laboratory would elevate the Air Force to

Gen Henry Viccellio Jr. approved the single-lab concept.

the same level and force-structure configuration as the Army and Navy, each of which had already consolidated its resources and political clout into one major laboratory.[15]

For a number of reasons, Viccellio immediately was very receptive to Paul's initial comments and explanations about the prospects of a single lab. During his first two years as commander of AFMC, Viccellio had formed some very definite opinions about the laboratories—not all of them positive. Firstly, he believed that the labs carried too much overhead—specifically, excess support people. Secondly, he felt very strongly that each of the labs was operating too independently. "The evidence that most drove me to that conclusion," according to Viccellio, "was the fact that when I went around on my initial visits to the labs, each of them spent quite a bit of time describing their marketing functions. In other words, rather than have a planning function, they were out drumming up business. So in the back of my mind I always had the feeling our lab structure wasn't well integrated and coordinated."[16]

Viccellio had other problems with the labs. It had come to his attention on numerous occasions that customers often had a difficult time identifying the right person or group of people in the laboratory structure to conduct the research the customer wanted. In many cases, customers became extremely frustrated because of the lack of a central subject-matter expert to go to for information and advice. Customers found it difficult to coordinate with two, three, or four laboratories responsible for performing bits and pieces of "very close to identical work." This duplication of effort and lack of synergy was of grave concern to Viccellio, and he wanted to fix that problem. He believed that one could cut out technical redundancy by creating a central plans-and-programs office located at the single lab's headquarters at Wright-Patterson. This new office would replace each laboratory's independent planning-and-programming shop that tended to serve its organization's narrow and most immediate interests. Instead of four separate labs going in four different directions, Viccellio wanted all four labs to work as one affiliated unit under one centralized plans-and-programs office to develop and move critical technologies needed to help operational commands meet their missions.[17]

Forced to reorganize the laboratory system because of the National Defense Authorization Act and *Vision 21*, Viccellio was not surprised when General Paul spoke to him on the airplane, suggesting the integration of all four labs under one central control. As Viccellio put it, "I was kind of in the mood for some initiatives along that line [consolidation]." After listening intently to what Paul had to say, he told Paul to "flesh it out" and give him some specifics as to what a new laboratory structure would look like. Viccellio also wanted Paul to begin identifying functions at each of the four labs that could be transferred into a consolidated lab headquarters as a way to reduce manpower slots across-the-board.[18]

Unlike Paul, General Viccellio had not grown up in the S&T community. A graduate of the Air Force Academy, Viccellio was a command pilot who had amassed over thirty-three hundred hours in fighter aircraft. First and foremost, he was operations oriented, having spent a large part of his career flying and maintaining a variety of aircraft. Over the years, he had developed a strong understanding of and appreciation for the role of logistics in the Air Force. When he moved from his position as commander of Air Education and Training Command to head up AFMC on 30 June 1995, he came with a healthy attitude that questioned what S&T could do for operational units. To him, the acquisition process was a complex and cumbersome experience, slow in transitioning the products of S&T to operators in the field. In short, deep down he wasn't sure how much all the investments in technology programs run by the labs were worth.[19]

Viccellio certainly recognized the value of research, but he also sensed that the current configuration of the labs was not the most efficient way to conduct business. In fact, he had been in favor of "closing one or two labs" as part of the BRAC 95 exercise. However, the BRAC commission rejected this plan, mainly because of political pressures exerted by congressional delegations to protect their local constituents from losing their jobs. So when all was said and done, Viccellio, as AFMC commander, was primed to offer some fundamental changes on running the labs more efficiently. Timing was the key. The appearance of the National Defense Authorization Act of 1996 and *Vision 21* in the spring of that year gave Viccellio and Paul the perfect opportunity and rationale to put their plan into operation to dramatically reform the organizational structure of Air Force laboratories. Furthermore, in the process they would be able to avoid and distance themselves from the politically sensitive issues of BRAC. More importantly, they would be able to accomplish the goal of laboratory reform in a low-keyed manner without the highly emotional public attention and political firestorm that BRAC had fueled.[20]

During the very early stages of their discussion in the late spring and early summer of 1996, Viccellio and Paul were very careful to keep their plans for the laboratory reorganization private. Neither was inclined to share his ideas with others at this point. They believed that if too many people became involved early on in the process of mapping out radical changes in the laboratory organization, their initial reaction might be to "kill the idea before it even had the chance to start." Further, both men realized that the Air Force did not want a lot of publicity on proposed lab reorganization until the secretary of the Air Force gave her final approval for a specific plan. Speculation about doing away with the four laboratories would have done more harm than good. Paul and Viccellio were extremely cautious at this juncture because if people began hearing about a single-lab proposal, they might prematurely and inaccurately jump to conclusions about the near- and long-term effects of closing down four labs. Naturally, people would be concerned about their jobs, careers, and future. In addition, politics was definitely a factor that could easily trigger congressional reaction in districts where the four current labs were located.[21]

Both Viccellio and Paul worried about possible interference from Congress that might impede any plans for establishing a single lab. If Congress became informed of even a tentative lab-reorganization plan, then the chances of moving forward quickly would most likely be slowed down. It seemed reasonable that congressional delegations would want to take a hard look at the specifics of any new laboratory plan that could affect the livelihoods of their constituents. Further, the congressional delegations would be out to actively lobby for their state to retain the new laboratory components in their geographic areas. Overall, Viccellio and Paul may have overestimated the resistance that Congress might use to prevent the formation of the single lab. After all, Congress had been the driving force for and a strong advocate of laboratory reform, as demonstrated by the passage of the National Defense Authorization Act of 1996. It was Congress that directed the labs to look at ways to consolidate and operate more efficiently. Congress also supported the downsizing of the military services, so it made little sense for Congress to go against its own policy. As it turned out, Congress, through its actions and desire for change, offered steady support rather than opposition to the Air Force's efforts to form one consolidated laboratory.[22]

Both Viccellio and Paul realized the potential repercussions and dangers associated with the premature release of information about proposing a single laboratory. Viccellio insisted that he and General Paul not discuss the plan for a consolidated lab with others during the very early conceptual phases. However, Viccellio made one important exception. Rather than springing such a radical proposal on Headquarters Air Force at some later date, Viccellio believed that from the very beginning, it was politically astute and absolutely essential that Darleen Druyun, the principal deputy assistant secretary of the Air Force for acquisition and management, be advised of the overall game plan. By bringing Druyun into the fold during the early concept phase of a single laboratory, Viccellio created an opportunity to test the waters at the higher levels of command. He wanted to elicit a reaction to the single-lab idea and get some kind of assurance that he would be backed up later. Viccellio made a special effort to keep in touch with Druyun on a regular basis and to keep her apprised of the initial plans for establishing a single lab and the thinking behind those plans. According to Viccellio, he and Paul "went up and briefed her on our concepts and again we kept it very close hold until we were ready to announce a steady plan and possible transition." The results of these contacts were positive. After listening to the pros and cons of moving toward a single lab, Druyun proved receptive to Viccellio's long-range plan and encouraged him to proceed. With her support, General Viccellio knew he was now in a better position to move ahead quickly with plans to consolidate four labs into one.[23]

At the same time Viccellio and Paul were weighing the potential options for the future of the laboratory system in response to the enactment of the National Defense Authorization Act of 1996, DOD was working on finalizing its report on lab reform due to the president by February 1996. As a result of NSTC's recommendations to the president in May 1995, Clinton directed DOD to furnish him a detailed plan and realistic schedule for downsizing the laboratories. Asking DOD to provide the specifics for reforming lab operations clearly demonstrated the executive branch's unyielding commitment to keep attention focused on the critical issue of lab management for the future. The message was that something had to be done—and fairly soon. This was an important step, but the report covering the details of laboratory reform that the president directed DOD to prepare for him by February 1996 never materialized.[24]

Circumstances overtook events. Congress became a second major player in lab reform with the passage of the National Defense Authorization Act of 1996. Ironically, this meant that there were now two studies under way at the same time addressing the same topic of laboratory reform—just the type of duplicative effort and waste of resources the president and Congress were trying to eliminate in DOD operations. To resolve this problem, the deputy secretary of defense sought clarification from the president and Congress. The outcome of all this was that the two studies already simultaneously under way—the DOD report to be put together as a result of NSTC's recommendations on laboratory restructuring (due to the president in February 1996) and the *Vision 21* DOD plan being prepared in response to the National Defense Authorization

Act of 1996—were to be combined into one *Vision 21* final report.[25]

Publication and delivery of the *Vision 21* report to Congress on 30 April 1996 did occur on schedule. As its name implied, the report proposed a vision or blueprint to create a new laboratory infrastructure that would reduce cost, eliminate unnecessary duplication of technical efforts, and maximize the efficiency and effectiveness of R&D programs leading to the production of reliable operational systems. For the vision to become a reality, the report identified three integrating "pillars" (fig. 1) that described what DOD had to do to successfully design and build a responsive laboratory infrastructure that would serve the nation's S&T needs for the future:

1. *Reduction* of current infrastructure costs with particular emphasis on the elimination of old, high-maintenance, and inefficient facilities while retaining critical capabilities for the future. Options would include reducing infrastructure costs of both laboratories and the T&E centers. One option would reflect reductions in both laboratory and T&E center infrastructure by at least 20 percent beyond BRAC 1995 by the year 2005.
2. *Restructuring*, to begin with intraservice restructuring, including business-process reengineering, with an emphasis on cross-service reliance.
3. *Revitalization* to modernize aged critical laboratories and T&E centers, with emphasis on technologies of the twenty-first century, cross-service sharing, improved efficiencies, and reduced cost of operation and maintenance.

According to DOD guidance, all three pillars were to be pursued simultaneously and with equal emphasis.[26]

Since each military service organized differently to meet its R&D responsibilities, DOD wanted to make sure that none of the S&T organizations were left out of the *Vision 21* evaluation process. Consequently, for the purposes of the *Vision 21* study, DOD defined a laboratory as any DOD activity conducting all or one of the following functions: S&T, engineering development, systems engineering, and engineering support of deployed materiel and its modernization. Moreover, if an organization received S&T funding in its budget for basic research, applied research,

Figure 1. Infrastructure Requirements for Defense Laboratories (From *Vision 21:* The Plan for 21st Century Laboratories and Test and Evaluation Centers of the Department of Defense [Washington, D.C.: Department of Defense, 30 April 1996], 1)

and advanced technology development (known as S&T budget activities), then that agency qualified as a laboratory. An organization did not have to be "named" a laboratory to fit the DOD definition. The term applied to research institutes as well as any other defense agencies performing research, development, engineering, and technical activities. The intent was to capture and investigate all DOD units that performed S&T so all could be integrated into whatever new laboratory system emerged from the *Vision 21* study.[27]

Based on this definition, the Office of the Secretary of Defense developed a list identifying 86 military organizations that qualified as laboratories for consideration under the *Vision 21* study. Twenty-nine belonged to the Army, 38 to the Navy, and 19 to the Air Force (table 2).

With Secretary of Defense William J. Perry naming the specific Air Force facilities engaged in S&T in the *Vision 21* report, it became absolutely clear that the Air Force had no choice other than devising its own internal laboratory-restructuring plan. To make the process work, the secretary laid out a timetable to ensure the accomplishment of specific milestones.

Table 2

Air Force "Laboratories" Identified for *Vision 21* Study

1. Armstrong Lab, Brooks AFB, Texas
2. Armstrong Lab, Wright-Patterson AFB, Ohio
3. Armstrong Lab, Mesa, Arizona
4. Human Systems Center, Brooks AFB (engineering functions)
5. Wright Lab, Wright-Patterson AFB
6. Wright Lab, Eglin AFB, Florida
7. Wright Lab, Tyndall AFB, Florida
8. Aeronautical Systems Center, Wright-Patterson AFB (engineering functions)
9. Aeronautical Systems Center, Eglin AFB (engineering functions)
10. Oklahoma City Air Logistics Center (non-depot-related engineering functions)
11. Ogden Air Logistics Center, Hill AFB, Utah (non-depot-related engineering functions)
12. Warner-Robins Air Logistics Center, Robins AFB, Georgia (non-depot-related engineering functions)
13. Phillips Lab, Kirtland AFB, New Mexico
14. Phillips Lab, Hanscom AFB, Massachusetts
15. Phillips Lab, Edwards AFB, California
16. Space and Missile Center, Los Angeles, California (engineering functions)
17. Rome Lab, Griffiss AFB, Rome, New York
18. Rome Lab, Hanscom AFB
19. Electronic Systems Center, Hanscom AFB

Source: Vision 21 Report, appendix E, 30 April 1996.

The game plan called for developing by 1 April 1998 a detailed five-year plan covering how the DOD laboratories would restructure. Secretary Perry would review and approve the plan and then submit it to the president on 1 July 1998 for his review and endorsement. The next step in October 2000 was to begin execution of the five-year plan. The mechanics of implementing the plan would take five years, with the completion date set for 1 October 2005. With this schedule on the table, each service now had only one option—to come up with a very precise plan on how it planned to reorganize.[28]

This came as no surprise to Viccellio and Paul, who already had been acutely aware of and heavily influenced by the very direct language of the 1996 National Defense Authorization Act, stating that labs would reorganize. Since their first encounter on the airplane, the two generals had engaged in several brainstorming sessions to discuss the lab issue. Their goal was not to work out the exact details of how the changes in laboratory structure would take place—there would be time for that later. Rather, these initial discussions were broad in scope and primarily looked at ways to make fundamental changes in lab operations across-the-board. The outcome of all this was swift and decisive. Both men had made up their minds in the June time frame—shortly after the release of the *Vision 21* report—that the best and most realistic course of action for the Air Force called for laying out plans to consolidate all of its S&T activities under a single laboratory.[29]

The meetings between Generals Viccellio and Paul that began in the spring and extended into the summer of 1996 were essentially informal but very productive sessions in Viccellio's office to exchange information and ponder various lab-reorganization options. For the first one-on-one meeting, Paul himself sketched out eight or 10 charts proposing a single lab. Rather than convening a group of his staff, instructing them to put some charts together on a single lab, and demanding that they adhere to those principles, Paul explained that things were moving quickly and that he just did not want a lot of people working on such a controversial and potentially explosive issue—at least not during these very early, speculative stages. "There wasn't time," according to General Paul, "so it was more of me proposing to General Viccellio, General Viccellio saying, 'Yeah, let's pursue this,' fleshing it out some more," until Paul and Viccellio reached a consensus.[30]

The consensus came quickly. One could argue that the earlier discussion on the airplane returning to Wright-Patterson really was the turning point, when both men first became sold on the idea of a consolidated lab. But that was not the case. They made no snap judgments on the plane. Instead, they prudently took additional time in the weeks that followed to think about the pros and cons involved in setting up one large laboratory. Only after presenting his charts proposing a single lab at their first meeting, Paul recalled, did General Viccellio officially give "the thumbs up."[31]

So the two generals had worked effectively together to reach a decision, but Paul fully realized that his role in this process was strictly that of an advisor deeply concerned about the future of Air Force S&T. Nevertheless, Paul's contribution was extremely significant because he had planted the seed and made the recommendation for a single lab. However, he did not make the final decision. Only General Viccellio had the authority to make the command-level decision to move toward a single lab. As Paul put it, "It was his [Viccellio's] call," and he made it. But even Viccellio did not have absolute authority. At some time in the future, he still had to go to the chief of staff of the Air Force and the secretary of the Air Force for their approval before he could begin to set the wheels in motion to establish a single lab. All this would take time and a great deal of preparation to spell out the reasons why a single lab would make sense.[32]

Viccellio realized that time was critical in making a decision on the single lab. He knew he had only a relatively short time to react to the National Defense Authorization Act and *Vision 21* and to come up with a sound plan by November 1996 for realigning the organizational structure of the four large laboratories under his command. He didn't waste time but faced this problem head-on, placing a great deal of faith in General Paul's input to help him make his final decision. Paul served as a steady influence by providing answers to Viccellio's initial questions, ranging from the size of a single-lab headquarters to the role of mission coordinators. Paul brought to Viccellio "a kind of conceptual schematic of how a single lab would look," including rough manpower numbers. Paul's proposal favored a centralized approach to lab management that would lead to a highly efficient operation with fewer people. This made sense to Viccellio, who believed that this strategy would best meet the consolidation goals of *Vision 21*.[33]

So all these factors operating together at the same time—the National Defense Authorization Act, *Vision 21*, and General Paul's input—combined to lead Viccellio to make his final decision to go with the single lab. Viccellio put it in perspective when he commented, "As soon as *Vision 21* began to take shape, there was no doubt in my mind that on the S&T side, lab consolidation to a single Air Force lab was going to be one of our major initiatives that we were going to propose to the Office of the Secretary of Defense." General Paul clearly recognized the implications of Viccellio's decision. Paul recalled that "it was a big step when General Viccellio said, 'Let's take this forward'—that was a very big step."[34]

Although Paul purposely did not consult with or advise his staff on the single-lab proposal during his initial rounds of private talks with Viccellio, he eventually did confide in and sought the advice of three of his closest associates. After the first couple of meetings with Viccellio, Paul shared his and Viccellio's commitment and enthusiasm for a single lab with Dr. Daniel, his deputy director, and Tim Dues, who ran the Plans and Programs Office (XP) at the Science and Technology Directorate. A third confidant drawn into the fold was Dr. Vince Russo, director of the Materials Lab, located at Wright-Patterson AFB. Russo, who had anchored the University of Rochester's offensive line in the early 1960s, would become a leading figure responsible for spearheading the drive to move the new single laboratory down the field. Paul personally selected Russo to be his S&T representative on the *Vision 21* committee chartered to develop the Air Force's plan for establishing its single lab. Paul, who had a great deal of confidence and respect for Russo's management style and abilities, later handpicked him to head the single-lab transition team in December 1996. In sum, Paul realized that the formation of one lab was simply much too big and complex an undertaking for any one person to handle. After Viccellio made the decision to pursue the single-lab option, even though it had not been officially announced yet in the spring of 1996, Paul quickly brought Daniel, Dues, and Russo in on the ground floor of this extremely important exercise.[35]

Paul viewed these men as the nucleus of his team that would make the single lab happen. More specifically, Paul counted on their expertise to make a smooth transition from the existing multilab configuration to one consolidated Air Force lab. They, more than any others who worked for him, thoroughly understood the big picture of how the labs operated on a day-to-day basis. Daniel, Dues, and Russo were the equivalent of the modern backroom politicians who could put the deals together, run interference for their boss, and influence and pressure competing factions throughout the laboratory system to compromise on the critical issues of restructuring.

Tim Dues, a key member of General Paul's inner circle, headed the Plans and Programs Office.

Dr. Vince Russo was a strong proponent of the plan for a consolidated laboratory.

Moreover, he trusted them. They had been around the block in the S&T business, and they were readily attuned to potential warning signs and obstacles at all levels that would have to be overcome before any single lab could be set up.[36]

In short, Paul depended more and more on Daniel, Dues, and Russo for their input on various aspects of lab restructuring during the formative stages, beginning in the early summer of 1996 and ending with the stand-up of the lab in October 1997. Even before that, Paul had had numerous conversations with all three to solicit their overall philosophy on lab reorganization. Daniel recalled, for example, that he and Paul had conducted many informal meetings in the small, private hallway between their adjoining offices. These discussions, which covered laboratory options, including a consolidated lab, periodically "came up from time to time" during the two years prior to *Vision 21*, usually prompted by projected budget and personnel cuts. As Daniel described it, these five- or 10-minute hallway discussions were part of an overall philosophy that "had been going on for some time . . . in little sound bites, but nevertheless inching along continuously between the two of us, but not discussed broadly with other folks." As Daniel interpreted events, it was clear to him that the thought of a single lab did not entail a singular event or one landmark decision when the light suddenly came on. Rather, the "thought evolved over a period of perhaps two years" prior to the commotion brought on by *Vision 21*.[37]

If Daniel, Dues, and Russo were to provide effective input, General Paul had to inform them as soon as possible about how the specifics of the proposed lab reorganization were unfolding at the highest levels of the Air Force. They needed to know the complete long- and short-range game plans if they were to play an active and convincing role in leading the charge for mustering support throughout the organization once the decision to set up one laboratory had been publicized.[38]

Publicizing the plan to move to a single lab would not take place for another five and one-half months. Formal announcement of the single-lab plan to all of General Paul's staff did not occur until after the Corona conference completed its work in October 1996 because, between May and October, General Paul, Dr. Daniel, Tim Dues, Dr. Russo, and a small, select group of other staff members needed time to help build a strong case outlining the rationale and transition to a single laboratory. They had to develop and support in more precise terms the advantages Paul and Viccellio had discussed on a global level. Doing so required expanding General Paul's inner circle by reaching down into the organization to get help from selected personnel to study and report on different aspects of the plan for establishing a consolidated lab. They undertook this time-consuming work solely to meet the immediate goal of developing input for presentation by General Viccellio at Corona so that he could obtain approval from the chief and secretary to proceed with the new laboratory plan.[39]

Notes

1. Maj Gen Richard R. Paul, interviewed by author, 2 March 1998.

2. Ibid.; Tim Dues, director, Plans and Programs Office, Science and Technology Directorate, interviewed by author, 2 and 12 March and 6 April 1998; and Dr. Vincent J. Russo, director, Materials and Manufacturing Directorate, Air Force Research Laboratory, interviewed by author, 4 February 1998.

3. Paul interview.

4. *National Defense Authorization Act for Fiscal Year 1996,* sec. 277, Public Law 104-106, 10 February 1996.

5. Ibid.

6. Maj Gen Richard R. Paul, director, Science and Technology, to Mr. Money, SAF/AQ, letter, subject: Defense Laboratory Consolidation, 12 February 1996.

7. Ibid. Air Force recommendations to BRAC 95 for closures/consolidations included moving Rome Lab to Hanscom AFB, Massachusetts; moving Armstrong Lab to Wright-Patterson AFB, Ohio; and cantoning Phillips Lab at Kirtland AFB, New Mexico, as a way to save money. BRAC, mainly because of political and fiscal considerations, rejected these Air Force proposals.

8. Paul interview.

9. *United States Air Force Biography: Major General Richard R. Paul;* on-line, Internet, January 1998, available from http://www.af.mil/news/biographies/paul_rr.html.

10. Ibid.

11. Dr. Robert Barthelemy, former director of the Training Systems Program Office, Air Force Materiel Command, interviewed by author, 6 February 1998; and Dr. Don Daniel, interviewed by author, 27 July 1998.

12. Barthelemy interview; and Daniel interview.

13. *United States Air Force Biography: Dr. Donald C. Daniel;* on-line, Internet, June 1997, available from http://www.af.mil/news/biographies/daniel_dc.html; Paul interview; and Col Dennis F. Markisello, vice commander, Air Force Research Laboratory, interviewed by author, 6 February 1998.

14. Markisello interview; Paul interview; and Daniel interview.

15. Paul interview; Gen Henry Viccellio Jr., USAF, Retired, former commander, Air Force Materiel Command, interviewed by author, 24 June 1998; and Gen Henry Viccellio Jr. to commander, Aeronautical Systems Center et al., letter, subject: Single Air Force Laboratory, 26 November 1996.

16. Viccellio interview.

17. Ibid.

18. Ibid.

19. *United States Air Force Biography: General Henry Viccellio Jr.;* on-line, Internet, August 1995, available from http://www.af.mil/news/biographies/viccelli_h.html; Lt Col Pat Nutz, AFRL/XPZ, interviewed by author, 4 February 1998; and Paul interview.

20. Nutz interview; and Paul interview.

21. Viccellio interview; and Paul interview.

22. Viccellio interview; and Paul interview.

23. Viccellio interview; Paul interview; and Dr. Sheila E. Widnall, former secretary of the Air Force, interviewed by author, 7 July 1999.

24. *Vision 21: The Plan for 21st Century Laboratories and Test and Evaluation Centers of the Department of Defense* (Washington, D.C.: Department of Defense, 30 April 1996); and Markisello interview.

25. *Vision 21;* and Markisello interview.

26. *Vision 21.*

27. Ibid.

28. Ibid.

29. Paul interview; and Viccellio interview.

30. Paul interview; and Viccellio interview.

31. Paul interview.

32. Ibid.

33. Viccellio interview.

34. Ibid.; and Paul interview.

35. Viccellio interview; Paul interview; and Russo interview.

36. Viccellio interview; Paul interview; Russo interview; Daniel interview; and Dues interviews.

37. Paul interview; and Daniel interview.

38. Paul interview; Daniel interview; and Dues interviews.

39. Paul interview.

Chapter 4

Overhauling Infrastructure

The National Defense Authorization Act for Fiscal Year 1996 and *Vision 21* were the two main actions initiated by Congress and DOD that caused the Air Force to commit fully to overhaul its laboratory infrastructure, which meant people as well as facilities. A day after the release of the *Vision 21* report, Undersecretary of Defense John White clarified the intent of the National Defense Authorization Act in a letter to all the military services: "The essential core work required of our laboratories and test centers to field the weapons systems of the future must be done without over-taxing the defense budget with unnecessary infrastructure." To achieve this objective, White directed that the plan to reorganize the laboratories had to consider options by which lab infrastructure could be "reduced by at least 20% by 2005."[1]

Manpower Reductions

Congress had not imposed a specific percentage in terms of how many personnel slots would be reduced from the Air Force rolls as part of the lab-restructuring effort. The legislation was very broad in its guidance to the military departments to consolidate into "as few laboratories . . . as is practical and possible." DOD wanted to include more precision in its guidance and therefore cited 20 percent. But even this percentage was not hard and fast. It was simply a starting point that DOD believed each military service and DOD collectively should strive to meet. Over the next few years, circumstances such as increases or decreases in annual budgets could cause this figure to fluctuate. Regardless of future budget changes or any other factors, Viccellio and Paul understood that the immediate task at hand entailed making significant personnel cuts.[2]

"One major reason for creating the new AFRL structure," according to General Paul, "was to streamline support and management, to maximize retention of scientists and engineers as we continue to downsize." From the very beginning, Paul realized that the definition of *streamlining* meant "a way to operate leaner [and] more efficiently with less overhead." Fashioning a leaner organization did not solely apply to reduction of the bricks-and-mortar infrastructure consisting of buildings, hardware, and lab facilities. To make progress, much of the downsizing would obviously come about by eliminating a substantial number of personnel slots. Unlike previous personnel exercises that targeted both scientists and staff, this time the lab reductions would come strictly out of the staff, where most of the overhead resided.[3]

Overhead would not go away voluntarily and thus became the driving issue at the highest levels. Since the inception of *Vision 21*, Vince Russo had been attending meetings at AFMC and Headquarters Air Force to keep abreast of the most up-to-date guidance for downsizing the labs. Most of the time, he worked closely with Alan Goldstayn, deputy director of the AFMC Plans and Programs Directorate, which reported directly to General Viccellio. Russo also routinely interacted with Blaise Durante, deputy assistant secretary for management policy and program integration, who served as the *Vision 21* lead point of contact for Secretary of the Air Force Sheila Widnall. Durante, Goldstayn, and Russo formed the core of the Air Force team responsible for planning and implementing *Vision 21*—the Army and Navy formed similar teams. All three service teams worked under the direction of Dr. Lance Davis, who worked for Secretary of Defense William Perry and coordinated the entire *Vision 21* effort out of DOD's Defense Research and Engineering Office in the Pentagon. Davis, a strong supporter of lab consolidation across the three military services, even favored closing some lab facilities. He envisioned the Army, Navy, and Air Force combining their resources and expertise to support specific technology programs at one geographic site.[4]

In many respects, Lance Davis was the key figure in the *Vision 21* process. He was the point man at the Director, Defense Research and Engineering (DDR&E) who orchestrated, monitored, and gave direction and guidance to Russo and others swept up by *Vision 21*. In turn, Russo's main job was to keep General Paul informed of all aspects and the

current status of the *Vision 21* process. A strong supporter of
the single-lab concept from the very start, Russo believed it
was feasible to downsize the laboratories. In his mind, the
laboratory system was broken, especially the composition of
the personnel force. As director of Wright Laboratory's Materials Directorate, Russo had become dismayed over the years as
he witnessed firsthand how the lab staff had grown out of
proportion. He was convinced that the organization had become top-heavy at the expense of the technical workforce: "I
hadn't hired a scientist or engineer in this damn place for
seven years—seven years!" However, Russo was also a realist
who, for better or worse, understood how slowly the wheels of
the government personnel system turned. Consequently, he
told General Paul that he "could not just whack off 20 percent" of the current assigned personnel. To reach 20 percent
savings in personnel would take BRAC-like authority. But
chances for another BRAC in the near future, according to
Russo, were extremely remote.[5]

A better plan involved achieving personnel savings by combining existing functions into fewer and larger organizational
elements as a means to help get rid of the overhead. Eliminating four separate lab commanders and their front-office staffs,
combining support functions such as the four lab-plans-and-
operations shops into one centralized office, and reducing the
number of tech directorates would all add up to lower overhead. To make this happen, however, required reorganizing
the entire laboratory structure. Russo insisted that no specific
number of personnel savings be given at the front end of the
Air Force *Vision 21* process. As he put it, "None of us knew."
One could arrive at those precise numbers only after reviewing
every position on each lab's unit-manning document to determine what jobs would remain under the new organization.
Russo also reasoned that it was not prudent for General Paul
to promise exactly how many personnel savings could be
made because those exact figures would remain unknown until they defined the new technology directorates. In general
terms, Russo advised Paul that "we can make a big step to-
wards that goal [20 percent reductions]." But several years
would pass before the emergence of more exact data showing
the extent of personnel savings.[6]

Col Dennis Markisello was one of General Paul's trusted advisors on the new lab.

Col Dennis Markisello, military deputy for the Science and Technology Directorate, joined Daniel, Dues, and Russo in the spring of 1996 as the fourth key figure brought into General Paul's inner planning cell to assist him in making the new lab a reality. Markisello had daily contact with Paul, and many of their discussions focused on cost savings derived from reducing personnel overhead. Markisello recalled that Paul's basic plan was to eliminate a good portion of the support layer of management as a cost-savings measure. Reduction of personnel was not something new to Paul and his staff. As Markisello put it, before *Vision 21* appeared on the scene, DOD continually pressured Paul's organization to downsize. Three important long-range personnel exercises designed to reduce the number of people assigned to the labs had already been under way for several years. One was the Dorn cuts. Another was the reductions imposed by the defense planning guidance (DPG). The third involved A-76 studies that systematically evaluated whether private-sector contractors could perform laboratory functions more efficiently and cheaply than civil servants.[7]

Dorn Cuts

The Dorn cuts were one of several critical personnel actions that significantly influenced the Air Force's decision to move to a single laboratory. Edwin Dorn, who served as the undersecretary of defense for personnel and readiness in the mid-1990s, played a prominent role in leading a determined effort to reduce DOD's civilian-manpower pool. Prior to coming to the Pentagon, Dorn had earned a PhD in political science from Yale and had held a variety of high-level positions at the

Brookings Institution and the US Department of Education. Because of his position and leadership role in directing major civilian-personnel reductions for the secretary of defense, these personnel actions that began in 1994 were dubbed the "Dorn cuts" throughout the military services.[8]

The origins of the Dorn cuts go back to the Bush administration's decision after Operation Desert Storm to reduce the active military forces by 30 percent. The goal was to gradually complete this drawdown by 1998. At the same time the active forces were moving ahead to implement hefty reductions in the military ranks, the Clinton administration in 1993 appointed John Deutch and Adm David E. Jeremiah, vice chairman of the Joint Chiefs of Staff, to cochair the Infrastructure Review Panel. This group would take a thorough look at reducing DOD's infrastructure and the effect that would have on the permanent workforce. One of the key aspects of this process was that DOD wanted the services to come up with solutions that would allow them to implement "voluntary" measures to cut back on the number of their civilian employees. Not surprisingly, all the services reported to DOD that they could not justify voluntarily downsizing their civilian workforce without seriously jeopardizing their missions.[9]

Also under way in 1993 was DOD's "Bottom-Up Review" (BUR), which examined the most prudent options for reducing the DOD labor force as the nation shifted away from a strategy designed to meet the Soviet global threat to one based on preventing aggression by regional powers. Emerging from BUR was a new strategy of engagement and enlargement that called for US forces working in concert with regional allies to fight and win two major regional conflicts occurring nearly simultaneously. In view of this changing mission profile, the report stated that one of its goals was to reduce the number of military personnel from a peak end-strength of 2.2 million in fiscal year (FY) 1987 to 1.4 million by FY 1999. The report recognized that a civilian drawdown would also occur but gave no specific numbers to indicate how extensive that reduction would be. However, the report mentioned that, like the plans for military and reserve separations, "plans for civilian separations will minimize involuntary departures. DoD intends to reach the civilian reduction level first by attrition, then by using the authorized buyout provisions recently passed by the Congress, and last, by involuntary separations." The report went on to point out that DOD would continue to adhere to its current restricted-hiring policy, replacing two civilian employees for every five employees who left civil service. This policy had begun in March 1991 and ended in the spring of FY 1994.[10]

Since the military services showed no inclination to voluntarily reduce civilians, Secretary of Defense Les Aspin convened a meeting with his deputy, William J. Perry, and Edwin Dorn to discuss alternatives. This meeting took place shortly before Aspin left his post on 3 February 1994. A turning point in the meeting came when Dorn presented a chart depicting the ebb and flow of the number of active duty force personnel and civilians assigned each year to DOD from 1950 to the present. Perry expressed particular interest and concern about this chart because it clearly verified dramatic reductions in the military forces during times of relative peace. But the dilemma, as Perry saw it, was that although the military took substantial reductions, the DOD civilian workforce—in some cases over the same time period—had not taken a proportional share of personnel reductions. If civilian jobs existed to support the military, then it seemed only logical that a proportional number of civilian cuts should follow military cuts. This was axiomatic to Perry's thinking, especially in light of the planned 30 percent military reduction through 1998.[11]

Perry was somewhat baffled. The military cut almost one-third of its active duty force from 1994 through 1998, but initially no plans existed to make proportional reductions in the civilian workforce. DOD projected that the civilian workforce would be "straight-lined" for the remainder of the 1990s, meaning that civilian authorizations would neither go up nor down. Perry did not like that scheme because it violated the historical pattern of corresponding cuts for both military and civilians during periods of downsizing. Consequently, he asked Dorn to go back and work the math to determine how many civilian positions would have to be eliminated so that reductions in the civilian force structure would be proportional to those planned for the military.[12]

Dorn came up with a plan that proposed spreading the reduction of civilian personnel numbers from 1994 through

2001. His blueprint for the future called for a 30 percent reduction over that eight-year period. If DOD adopted this prescription, he claimed that the projected DOD civilian figures for 2001 would strike the right balance with the reduced military personnel numbers scheduled for the end of 1998. He presented his blueprint in the spring of 1994 to Perry, who had replaced Aspin as secretary of defense. Perry liked the plan and told Dorn to issue instructions to get the process under way. On 2 June 1994, Dorn sent a letter to all the secretaries of the military departments specifying year-by-year projections of how many civilian positions each service had to remove. This became known as the infamous "Dorn memo" that directed each service to reduce 4 percent of its civilian workforce each year from 1994 through 1999. The year 2000 would see the imposition of a 3 percent civilian reduction, followed by a 2 percent reduction in 2001. Following this schedule and taking the prescribed cuts each year meant that by the end of 2001, DOD would diminish civilian positions by almost 30 percent.[13]

The percentages of four, three, and two were selected primarily because any number higher than 4 percent per year would trigger a major reduction in force (RIF)—something DOD wanted to avoid because of its extreme expense and political unpopularity. In addition, RIFs of all types demoralized the entire workforce. DOD realized that it had to manage the downsizing as efficiently as possible without sacrificing the mission in the process. As an integral part of this process, DOD also wanted to dispel the notion that it was a large and insensitive institution. If DOD truly believed that people were its most important resource, then the Pentagon had to practice what it preached and convey to its civilian workforce that it would reach the reduced numbers humanely as well as efficiently. Distributing the civilian cuts over eight years would lessen the stress and negative impact on both the overall workforce and specific individuals affected by the drawdown. With relatively small reductions (2 to 4 percent) each year, the workforce could absorb a significant portion of the personnel losses by normal attrition. DOD could further diminish part of the balance through special personnel initiatives such as incentive bonuses to encourage early outs, job transfers, voluntary early retirement, and removal of vacant positions from the books. Hopefully, all this would minimize the number of positions identified for a RIF.[14]

The Dorn cuts had an immediate impact on the operations of the four laboratories. First of all, there was to be no further debate over whether or not DOD would make substantial civilian cuts from 1994 through 2001. The secretary of defense had made that decision, and the Dorn cuts would implement it. The Dorn procedure had put civilian-reduction percentages in place for each year, and the AFMC commander now had to ensure that all his subordinate units, including the four laboratories, complied with Dorn guidance and took their fair share of the cuts. However, the laboratories would experience complications.[15]

In 1994 each of the four laboratories reported to a different system product center because each laboratory had a particular R&D area of expertise best suited to support a specific operational requirement of the Air Force. For example, Phillips Lab in Albuquerque worked with and reported directly to the Space and Missile Systems Center in Los Angeles. One of the lab's primary responsibilities was to advance new technologies that could be moved and applied to space systems developed and prepared for launch by the product center in Los Angeles. As part of this arrangement, the center's commander had complete control over all personnel resources, affording him a great deal of leverage in terms of determining which organizations would take the bulk of the cuts.[16]

General Paul found himself boxed in a corner when it came to dealing with personnel issues. As the Air Force's executive officer of technology, he had authority and control over all four lab budgets. But he had almost no say in the deletion of personnel slots. Each product center commander decided, under the umbrella of his center's unit-manning document, which positions would remain and which would go. General Paul could shift dollars between labs and program areas, but he could not move even one person from one lab to another or to his directorate without negotiating with one or more product-center commanders. This system did not make much sense to him. People were indeed an important resource, and Paul reasoned that basic "Management 101" and good business practices made it obvious that a single manager should remain accountable for all the organization's resources, including funding and personnel. Without that authority and control, commanders of organizations would operate with one hand tied behind their backs, while at the same time they were expected to accomplish the mission in the most expeditious and proficient manner possible. Establishing a single lab with one commander would go a long way toward solving that problem.[17]

Defense Planning Guidance: More Personnel Reductions

The decision to begin implementing the Dorn cuts in the summer of 1994 was one of the first indicators of what would follow over the long term in the area of eliminating positions in DOD and the Air Force. A severer blow delivered to the overall declining personnel picture came in the form of May 1995's DPG, a planning document prepared by the Office of the Secretary of Defense and released before each budget cycle. This document focused on identifying what programs and how many people each DOD organization would need to run these programs to meet the department's requirements in FY 1997. Budget officials used the information in the 1995 DPG as building blocks to prepare the Program Objective Memorandum for FY 1997, which included budget recommendations to ensure that moneys for programs and salaries would be in place so that each military unit would have adequate resources to complete its missions.[18]

The DPG's "overarching plan" for S&T programs focused on developing superior technology for creating and maintaining an effective and affordable military capability. More specifically, this strategy embraced the idea that future military systems introduced into the inventory needed to be designed and built to achieve lower acquisition and life-cycle-ownership costs. Reduced costs applied not only to new hardware. One could also realize savings by reducing the number of people required to operate, maintain, and support these weapon systems. Likewise, the guidance directed DOD's S&T organizations to spend their money more wisely by exploiting commercially available technology, subsystems, and components for developing state-of-the-art systems tailored to meet very specialized military missions.[19]

Under the "Downsizing and Re-engineering Initiatives" section of the DPG, instructions described the quantitative nature of the drawdown for the three military services. At the heart of the matter was the proclamation that "the services should continue to reduce the aggregate number of full-time equivalents [jobs] RDT&E [research, development, test, and evaluation] activities." To add teeth to this broad policy statement, the DPG laid out very precise number goals. For the Air Force as well as the other two services, the DPG clearly outlined DOD plans that called for "reductions by FY 2001 of at least 35 percent from the aggregate peak personnel levels of each Military Department." The Air Force and other services had to cut their forces by over one-third in less than five years—a staggering number.[20]

However, the burden of the 35 percent figure was somewhat tempered because the 30 percent Dorn cuts already under way would count as part of the larger DPG reduction. In effect, the DPG added another 5 percent reduction to the Dorn cuts. Although on the surface, 5 percent did not appear extraordinarily large, in the real world it amounted to a big increase, considering the fact that as of May 1995 the Air Force had reduced by only 11 percent, leaving a balance of 19 percent to reach the level established by the Dorn cuts. Because of the extra 5 percent imposed by the DPG, the Air Force now had to take a 24 percent reduction (instead of the 19 percent) to meet DPG's goal of a 35 percent reduction. The extra 5 percent was substantial because it targeted an additional 1,140 jobs for elimination.[21]

The "peak personnel levels" referred to in the DPG were defined as the number of people assigned to each service as of 1991. In that year, the Air Force had 22,800 assigned to its RDT&E workforce, so it had to lose eight thousand people to meet its 35 percent reduction quota—something that would not happen overnight but would gradually occur over the next few years. As pointed out before, by May 1995 the Air Force—through the Dorn cuts and other personnel exercises—had already whittled its labor force by 11 percent, based on the FY 1991 personnel numbers. This translated to twenty-four hundred R&D slots that the Air Force had removed from the personnel rolls, leaving a balance of fifty-six hundred positions that still had to come off the books over the next six years to ensure compliance with the DPG. Based on these figures, the Air Force would have to eliminate almost a thousand jobs a year over the next six years—an unpleasant challenge for General Paul because it meant that large numbers of laboratory people would likely lose jobs that they had held for 10 or 20 years.[22]

Paul had always intended to make the drawdown of personnel as painless as possible, taking the position that a RIF would be his last choice as a means of trimming the workforce. He preferred to lose people through the normal attrition process, whereby employees voluntarily left the workforce through retiring or by moving on to new jobs in other government agencies or private industry. He also planned to offer monetary incentives to entice people to consider early retirement or to separate from civil service outright. Deep down, General Paul knew that normal attrition would not take care of the entire problem. A realist, he recognized that he eventually would have to invoke a RIF to accomplish the personnel reductions mandated by DOD. What bothered Paul the most was the prospect of dismissing people who were high-quality workers. Under RIF guidelines, no matter how well individuals might perform their duties, they remained targets for removal, especially if they were junior in rank or new to the organization (last in, first out). Financial resources simply no longer existed to support a large laboratory workforce in view of perceived global conditions that focused on regional conflicts rather than the traditional, massive Soviet threat.[23]

In the grand scheme of planning for the future of S&T to meet declining personnel requirements associated with the Air Force's changing mission, the DPG was the straw that broke the camel's back. It was a turning point that convinced General Paul of the inevitability of combining the various pieces of the Air Force's R&D organizations into a single lab. Harboring no doubt that the 35 percent DPG reduction represented "a huge cut," Paul made it one of the primary reasons for deciding that a single lab made the best sense.[24]

Looking back on the events of 1995, General Paul reflected on the importance and sway that the DPG had had on his thinking: "Like I said, that [DPG] is the dynamic more than anything to me that said we just can't continue to keep the old—the current organization—and downsize in place." He pointed out that because the four labs—including their divisions, branches, and staffs—numbered too many people, he had to find a way, as he described it, to "flatten" the current organization. In the process, Paul believed he had an opportunity to simultaneously accomplish other objectives, "like consolidating resources [staff and support functions] and reducing fragmentation [similar technology efforts taking place at different locations]. Those were kind of secondary goals—as long as we were going to streamline, why not try to take care of those at the same time?" The Air Force had to create a single laboratory to "see if we could get enough organizational efficiency so we could keep doing our missions but deal with a smaller organization." If reorganization did not occur, Paul predicted that he would have to reduce some missions. Because of personnel reductions that would definitely occur—and in some cases, would target scientists and engineers as technical programs were eliminated—some missions would have inadequate manpower. Consequently, those missions were strong candidates for removal. That was the dilemma Paul faced.[25]

A-76 Process: Contracts and Privatization

General Paul constantly felt the pressure on two fronts with the Dorn and DPG requirements. At the same time, he also had to contend with the BUR finding that included a section on infrastructure which affected the future of DOD laboratories. Although the discussion of this topic was rather brief, the message was clear: DOD would actively pursue a long-term process to reduce and streamline its infrastructure to achieve cost savings by increased use of privatization to take over selected government functions. Other recommendations included "consolidation of functions" and the "implementation of better business practices" to DOD operations. In essence, DOD had begun issuing strongly worded guidance encouraging military organizations—including the laboratories—to take a hard look at hiring contractors to replace certain civil servants and military personnel who would be let go as part of the overall downsizing process.[26]

With more and more pressure on government to downsize the workforce, the question of whether it was more economical to hire contractors to replace government workers in selected areas became a much bigger issue in 1996. The "A-76 process" was the method used to determine whether governmental work activities should be performed under contract, using commercial sources, or in-house, using government facilities and personnel. One can trace the origins of this process to the Budget and Accounting Act of 1921, which supported the policy that the government can rely on products and services from the commercial sector. A-76 guidance first appeared in Bureau of the Budget bulletins in 1955. In 1966 the Office of Management and Budget (OMB) published Circular A-76, revised in 1967 and again in 1979. In March 1996, OMB published a new version of Circular A-76's *Revised Supplemental Handbook* that provided updated guidance and spelled out point-by-point procedures for determining if it was more economical to hire commercial businesses to "supply the products and services the Government needs."[27]

The government's policy did not call for automatically turning over government workers' jobs to the private sector. Rather, the heart of the A-76 process involved fostering competition between the government organization and any private company that claimed it could do the job cheaper. Before any final decision, each party had to prepare a comprehensive cost-comparison estimate reflecting the expense of performing jobs classified as "not inherently governmental functions." These cost calculations depended upon a variety of variables—number of workers required to perform the function, total salaries and benefits, condition of facilities and equipment needed, maintenance and repair, and more.[28]

If a private agency underbid the government organization, then it could receive a contract to perform that job. The obvious benefit was that the government could then remove the government workers who previously performed that duty from the books, thereby contributing to the reduction of government workers on the payroll. Another option was that after reviewing the performance of a particular job function, the government could recommend reducing the percentage of government workers performing that job and thereby calculate a lower cost than the one bid by the contractor. In that case, the government would profit by keeping costs down and retaining a portion of its original workforce. At the same time, the government would comply with the downsizing goals by eliminating a certain percentage of government workers no longer required.[29]

Although A-76 appeared to provide some very tangible benefits in terms of costs and personnel reductions, one could easily overlook some fundamental drawbacks to this process. To many people, A-76 was a tortuous ordeal. Firstly, government preparation of a lengthy and detailed cost analysis proved extremely time-consuming and very costly in terms of total man-hours and dollars. Employees' efforts to devise detailed work-performance statements, collect and interpret a maze of financial data, and meet other convoluted A-76 requirements took up large portions of time normally devoted to their daily jobs. Consequently, in many cases the work they were hired to do suffered. This situation was exacerbated by the fact that the A-76 process often took more than a year to complete. Secondly, a lower contractor bid did not always guarantee the performance of *quality* work since A-76 focused mainly on reducing costs and government personnel. However, no accurate measuring stick existed for reliably determining ahead of time the level of quality work that a contractor would perform. If quality went down, then one could persuasively argue that keeping the original government workers might have been more cost-effective and efficient. Moreover, once contractors were on board, what would prevent them from significantly increasing the price of their contract two or three years down the road? Thirdly, A-76 was extremely demoralizing, appearing to be an insensitive and clinical process that produced a tremendous amount of stress on workers who prepared cost-analysis data that could very likely cost them their jobs.

General Paul and his staff planned to use the A-76 process as one means of achieving the personnel drawdown that applied to all four laboratories. In February 1996, a total of 540 positions across the four labs were identified as eligible to undergo A-76 evaluation. Paul realized that not all positions eligible for A-76 would convert to contractor work, but, based on projected personnel needs over the next few years, he estimated that A-76 might eliminate 359 of the 540 positions identified. In reality, however, one could not hope for quick results or large reductions in laboratory personnel—at least in 1996—because of the snail's pace of the start-to-finish A-76 assessment. From his vantage point, Dr. Daniel believed that "A-76 was not that big of a deal," especially when compared to the impact of the Dorn cuts.[30]

DOD's personnel figures confirmed Daniel's intuitions about the contribution of A-76. In a briefing to the Defense Science Board, Diane Disney, who tracked personnel trends for DOD, admitted that in the mid-1990s, "privatization has not been a major part of downsizing." Although the highest levels of government had addressed ideas about the benefits of privatization for some time, they had little to show in terms of meaningful personnel reductions. For example, in all of DOD from FY 1993 to FY 1996, a total of 19 A-76 studies had resulted in the separation of only 192 civilian employees. In spite of this relatively poor showing, DOD continued to take the position that A-76 would become "the dominant factor for the future." Although the department had not vigorously pursued contracting out, which had not resulted in dramatic personnel reductions in the past, it recognized that it would have to endorse a policy of privatization to meet future personnel and budget goals. DOD's turnaround made sense if one believed that future budgets would be slashed and that fewer employees would be needed to perform the new missions of the twenty-first century. Only time would tell if this would become a realistic approach for making significant cuts in the overall personnel picture.[31]

Laboratory Personnel Profile: A Downward Trend

The years immediately preceding the stand-up of the Air Force Research Laboratory in 1997 significantly affected the personnel profile of the new lab. Even to the most casual observer, the first half of the 1990s portended fundamental change for the future. In keeping with the Clinton administration's policy of downsizing the military, manpower trends within each of the four Air Force laboratories revealed a continual decline in personnel numbers throughout the 1990s. DOD reform measures embodied in the Dorn cuts and the DPG convinced General Paul that consolidating civilian and military jobs under one laboratory was the best way to reach personnel-reduction goals.[32]

Looking back at the history of laboratory manpower showed that the laboratory organization—consisting of the four labs plus the Science and Technology Directorate (AFMC/ST), the Air Force Office of Scientific Research (AFOSR), and the Technical Transition Office (TTO)—had been consistent from year to year in complying with DOD's downsizing policy. From 1990 through 1996, total laboratory-authorized manpower positions decreased from 8,480 to 7,226—a loss of 1,254 slots, representing a reduction of 15 percent (table 3). By the end of FY 1997, the total manpower number had dropped to 7,091.[33]

However, this figure from FY 1997 was misleading because 511 Program 8 positions assigned to Armstrong Laboratory at Brooks AFB in San Antonio were transferred out of the laboratory and assigned to the Human Systems Center. Program 8 jobs included veterinarians, drug testers, medical technicians, and so forth, assigned to Armstrong Lab. These jobs were important, but in times of downsizing when reductions had to be made, Program 8 positions received a lower retention priority. General Paul and others did not believe that these 511 jobs fit the strict definition of pure S&T positions needed to carry out the laboratory's main mission of R&D. Consequently, the 511 Program 8 positions were moved off Armstrong Laboratory's unit-manning document and transferred to the Human Systems Center, also located at Brooks AFB. The problem, however, was that these positions were "moved" from one organization to another rather than deleted from the personnel books. Although these jobs no longer remained laboratory assets, the Air Force still had to pay the salaries and provide support services to the 511 people who now appeared on the Human Systems Center's unit-manning document.[34]

Table 3

**Total Laboratory Manpower
(Fiscal Years 1989–2001)**

Fourth-Quarter Data	FY89	FY90	FY91	FY92	FY93	FY94	FY95	FY96	FY97	FY98	FY99	FY00	FY01
AFMC/ST	0*	0*	101	114	108	104	97	96					
AFOSR	0*	208	209	222	217	216	146	146					
TTO	0*						41	37					
Armstrong	0*	2,010	1,660	1,553	1,556	1,629	1,530	1,526					
Phillips	0*	2,092	2,295	2,169	2,059	1,774	1,750	1,773					
Rome	0*	1,239	1,206	1,128	1,039	992	1,104	1,088					
Wright	0*	2,931	2,903	2,684	2,770	2,781	2,599	2,560					
AFRL									6,580**	6,330	6,259	6,183	6,144
Total		8,480	8,374	7,870	7,749	7,496	7,267	7,226	6,580**	6,330	6,259	6,183	6,144

*Accurate data not available
**511 Program 8 positions cut from AFRL

Source: AFRL Total Manning, Fourth Quarter, End of Fiscal Year Unit-Manning Document.

All this meant that General Paul's organization did not receive credit for moving the 511 positions to another Air Force
organization because the jobs did not go away. To receive
credit for personnel reductions under Dorn and DPG, positions had to be "eliminated" from the Air Force. That had not
happened. Consequently, when the single lab stood up in October 1997, the manpower numbers showed 6,580 authorized
positions (see table 3). In reality, the 511 positions had to be
added to the 6,580 number, resulting in a new total of 7,091,
of which 73 percent were civilians and 27 percent were military.
Although counting the 511 positions in this way might seem
unfair, it was the accepted accounting method used by DOD
to calculate personnel-reduction credits.[35]

In the midst of this somewhat confusing and changing personnel picture, one could argue that the 6,580 number cited
for 1997 was more accurate than the 7,091 figure. Realistically, the single lab had only 6,580 assigned positions upon its
official establishment in 1997. Calculating personnel losses
using the 6,580 number meant that a reduction of nineteen
hundred positions (8,480 minus 6,580) occurred from FY 1990
through FY 1997—a 22 percent decline. As mentioned above,
the decline amounted to 15 percent if one counted the 511
Program 8 positions. Either way, since 1990 the emerging general pattern showed a steady loss of manpower at the lab. All
signs indicated that this trend would accelerate in the future.[36]

Implementation of the Dorn cuts and DPG had both an
immediate and a long-term effect on downsizing the laboratory
workforce. But trying to gauge precise personnel numbers was
not always an easy task for a couple of reasons. Firstly, collecting accurate numbers on total authorized positions from
1989 forward proved difficult simply because all the data was
not available or had to be pieced together by combining information from a variety of source documents. Secondly, identifying vacant positions (likely candidates for elimination) on the
unit-manning document was also difficult because those positions underwent constant change—at least for the current
year. Although some vacant positions might be filled, other
positions might become vacant through retirements, changes
of jobs, resignations, and so forth. Furthermore, the fact that
individuals stated they would retire, change jobs, or resign did

not prevent them from changing their minds at the last minute, reducing the certainty in calculating an accurate list of projected vacancies. So the number of vacancies throughout the laboratory system changed almost daily, making it tougher to get a firm grip on the exact number of vacant slots near the top of the list to consider for elimination as part of the long-term game plan to downsize.[37]

General Paul knew that once the 35 percent reduction plan for laboratory personnel was finalized in May 1995 with the release of the DPG, he had to begin developing a strategy to meet that goal. One of the first steps was to establish a credible personnel baseline number to calculate what the lab would have to get down to in terms of authorized personnel positions by 2001. In a reduction-strategy briefing in February 1996, General Paul used FY 91 with a baseline of 8,015 to project future personnel losses. Subtracting 35 percent from the baseline left a remainder of 5,210 positions by the year 2001. This was a first cut at trying to determine where the laboratory would have to be in 2001 to comply with DPG's 35 percent directive (fig. 2).

Defining reliable personnel numbers to meet goals five years in the future was no easy task. Manpower numbers always seemed a moving target because changes and inputs to the

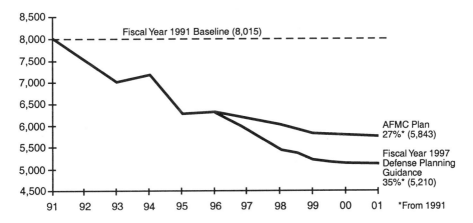

Figure 2. Air Force S&T Manpower Reductions, Program 6 (From briefing, Maj Gen Richard R. Paul, director, Science and Technology, Headquarters Air Force Materiel Command, subject: S&T Manpower Reduction Strategy [FY 97–01], 16 February 1996)

system occurred daily—even hour to hour. Accordingly, the number of assigned positions underwent continual revision to answer the most immediate question at hand. For example, the laboratory goal of 5,210 authorized slots for 2001 (see fig. 2) changed as plans to reorganize began to unfold throughout 1996. As more and more historical personnel data was collected and used, General Paul's staff built an accurate set of manpower numbers for FY 1989 in lieu of FY 1991. By using 1989 as the starting point—the year DPG originally designated as the "peak year"—to base manpower reductions, one could change the final manpower goal from 5,210 to 5,507 authorized positions. This new number came from taking away 35 percent of the FY 1989 baseline of 8,493 (fig. 3).

Fine-tuning manpower numbers with exacting precision for several years in the future was not the most pressing concern for reorganizing the laboratory. The important issue involved taking action to implement the spirit and law of the DPG that would lead to a smaller and more efficient workforce. Planning

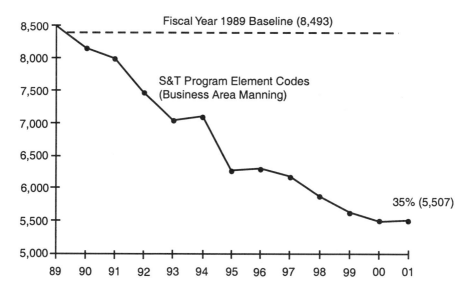

Figure 3. **Manpower Trends** (From briefing, Col Mike Pepin, AFRL deputy director, Plans and Programs, subject: AFRL Manpower Report, FY 89–FY 01, 20 March 1998; and fax, Jan Moore, AFRL/XP, to author, subject: S&T Manpower, 17 March 1998, with attached briefing chart, "S&T Manpower and Budget Trend," 23 January 1998)

and moving in the right direction, along with starting to identify and remove authorized positions from the unit-manning
document, comprised a critical first step that DOD leaders
wanted to see happen. However, they realized that projected
personnel-reduction numbers varied, depending on the latest
available manpower data for past years. Manpower was certainly
a core issue that the reorganizers had to deal with, but General
Viccellio sensed the urgency of the bigger picture. Primarily, he
wished to avoid delay in initiating the process of establishing a
single lab. He expected and depended upon General Paul to
carry out this exercise in the most expeditious manner possible.
Consequently, during the early stages of planning for the new
lab, he did not overly concern himself with defining and defend-
ing a specific 35 percent reduction number.[38]

Viccellio served as the decision maker who articulated the
general guidance that armed General Paul with the authority
to make the single lab happen. Viccellio did not want to get
bogged down in the details of personnel numbers. Although
the final 35 percent manpower number initially varied from
5,209 to 5,507 and then settled somewhere between fifty-five
hundred and six thousand by the beginning of FY 1997, Viccellio showed no inclination to name a specific number. General Paul knew this, noting that Viccellio "never put a goal in
writing" as regards long-term manpower numbers. However,
in his numerous meetings and interactions with Viccellio in
the spring of 1996, Paul had come to believe that Viccellio
thought "we would even go down below the 5,500 point. His
[Viccellio's] hope was that with the single lab that it would be
efficient enough that maybe there were another couple of hundred people beyond that."[39]

Many reasons led them to avoid devoting an excessive
amount of time and energy to perfecting the goal of achieving
a 35 percent reduction in manpower for 2001. Firstly, it was
up to Air Force Materiel Command to show it was on the
correct heading for achieving the 35 percent reduction. AFMC
had the responsibility of keeping track of laboratory personnel
cuts, but no outside office or agency was involved in the details of keeping an account of the exact number of cuts. Eventually, the command would have to report its progress periodically up the chain. Any heading checks, realignments, or

accelerated personnel cuts would occur and be negotiated at that time. In addition, Congress might enact legislation in 1997 that would later cancel, revise, or increase or decrease the cuts imposed by the 1995 DPG. In that situation, the 35 percent manpower number for 2001 would change.[40]

One of General Paul's major concerns in setting up the single lab was the selection process for determining the types of positions to give up as part of the overall downsizing effort. Higher authorities provided no guidance about the types of positions to eliminate during any of the personnel-reduction exercises. Higher headquarters simply passed on specifying how many reductions each organization had to make annually. Identification of specific job categories was left up to the leadership of the organization affected by the personnel cuts. During the first few years with the Dorn cuts, for example, laboratory cuts came from a variety of positions, including support staff and middle management as well as S&T job series. All of these work sectors contributed to the mandatory personnel drawdown.[41]

General Paul especially wished to do something to reverse the trend of cutting so deeply into science and engineering positions. After all, the heart and soul of the laboratory organization were the scientists and engineers who performed R&D. From 1994 to 1996, the Dorn cuts had consumed too many science and engineering positions. As Paul explained the mechanics of the reduction process, "We were taking support and researchers almost in an equal proportional number." At that point, one accounted for most of the personnel losses by giving up vacancies or through retirements and job transfers. Eventually, however, vacancies would run out. Although from 1994 to 1996, no one had to face a reduction in force, Paul anticipated the inevitability of RIFs in the future. But after expending all the vacancies, he would have to have a new system in place to ensure the protection of scientific and engineering jobs. Simply put, the erosion of the scientific and technical workforce could not continue if one expected the laboratory to accomplish its mission. General Paul summed it up best when he commented, "Without reorganizing, we were going to have to take a lot of our cuts from scientists and engineers."[42]

When *Vision 21* came along in 1996 advocating laboratory consolidation, it proved a timely and favorable opportunity and mechanism for preserving scientific and engineering positions within the laboratory system. General Paul believed that the decision to reduce the four labs to one would satisfy the consolidation intent of *Vision 21*. But it also would have other important benefits. Reorganizing into one lab would provide solid justification for removing overhead positions that previously existed in four separate laboratories. The thinking was that, instead of maintaining four separate command sections and their staffs, a single lab would require only one command section and a consolidated staff consisting of fewer people than the four collective staffs. Reorganizing in this manner meant that future personnel cuts would come exclusively from overhead positions because one lab would not need as many support and management people. Thus, scientific and engineering positions would be protected to ensure that R&D mission performance did not suffer. The only exception to this policy would occur if the Air Force decided to do away with a specific technology program of lesser relevance to users that it could not fund because of budget cuts. In that case, science and engineering positions would be dropped from the unit-manning document because the technical program these positions supported would no longer exist.[43]

Predictions of budget cuts influenced the speed of removing personnel positions under the new laboratory realignment. For example, from 1997 through projections for 2001, the lab's balance of 453 positions had to be eliminated as part of its remaining Dorn allocations. In 1997 the lab gave up 280 positions, leaving only 173 to be abolished over the next four years. Why was the laboratory in such a rush to get rid of personnel slots so much in advance of the 2001 deadline?[44]

The answer is that the lab wanted to act quickly because of projected budget cuts. If budgets decreased, then it would have less money available to pay civilian salaries. As Paul explained, "It is in our interest, as long as we are going to lose them [civilian positions] anyway, to lose them earlier; we don't need as many support people. The longer they are on the payroll, the longer we have to pay them; when we have to pay them, that is less money to do technology." So the desire to get civilian positions off the books quickly was in step with the idea that the single laboratory had to create a "flatter" civilian workforce and prepare as early as possible to deal with future budget cuts that would prevent the paying of all civilian salaries.[45]

Pressures brought on by personnel reductions, especially the requirements imposed by DPG and the Dorn cuts, contributed significantly to the decision to go with a single lab. Viccellio, Paul, and others in the Air Force hierarchy concluded that formation of a single laboratory represented the best way to reduce the number of civilians. As the consolidated lab came together, it would need only a relatively small staff and management force. Hence, creation of the single lab offered the rationale to reduce civilian positions and at the same time make the organization leaner and more efficient by centralizing its operations. A single lab would be more efficient because it would no longer need the "integrating function" of the AFMC/ST staff. The single lab commander and his staff now would perform that function.

Notes

1. John P. White, undersecretary of defense, to secretaries of the military departments, letter, subject: Plan for Consolidation of Defense Laboratories and Test and Evaluation Centers, 1 May 1996.

2. *National Defense Authorization Act for Fiscal Year 1996*, sec. 277, Public Law 104-106, 10 February 1996. Dr. Daniel believed that Paul Kaminski, undersecretary of defense for acquisition and technology, proposed the 20 percent figure.

3. Maj Gen Richard R. Paul, interviewed by author, 2 March 1998.

4. Dr. Vincent J. Russo, interviewed by author, 4 February 1998; and Brig Gen Michael C. Kostelnik, director of plans, to ALHQCTR/CC et al., letter, subject: *Vision 21* Laboratory and Test and Evaluation Consolidation Study, 7 August 1996.

5. Russo interview.

6. Ibid.; and Col Dennis F. Markisello, interviewed by author, 6 February 1998.

7. Markisello interview.

8. Dr. Edwin Dorn, dean, Lyndon B. Johnson School of Public Affairs, University of Texas at Austin, interviewed by author, 22 December 1998; and biography, Edwin Dorn, University of Texas, December 1998.

9. Dorn interview.

10. Les Aspin, *Report on the Bottom-Up Review* (Washington, D.C.: Department of Defense, October 1993), 81–84; and idem, *The Bottom-Up Review:*

Forces for a New Era (Washington, D.C.: Department of Defense, 1 September 1993).

11. Dorn interview.

12. Ibid.; and Linda Gileau, Office of the Undersecretary of Defense for Personnel and Readiness, Requirements Directorate, interviewed by author, 6 January 1999.

13. Edwin Dorn, undersecretary of defense for personnel and readiness, to secretaries of the military departments, letter, subject: DOD Civilian Resource Guidance—FY 1994–2001, 2 June 1994; and Gileau interview.

14. Dorn interview; and Gileau interview.

15. Tim Dues, interviewed by author, 2 March 1998.

16. Ibid. The other three labs had similar ties with their product centers. Armstrong Lab reported to the Human Resource Center at Brooks AFB, Tex.; Rome Lab reported to the Electronic Systems Center at Hanscom AFB, Mass.; and Wright Lab reported to the Aeronautical Systems Center at Wright-Patterson AFB, Ohio.

17. Paul interview.

18. Tim Dues, interviewed by author, 6 April 1998.

19. Dan McDermott, AFRL/XPP, to author, fax (with attached excerpt from FY 1997 DPG, 9 May 1995), subject: DPG, 23 June 1998.

20. Dan McDermott, AFRL/XPP, to author, fax (with attached excerpt from FY 1997 DPG, 9 May 1995, 64), subject: DPG, 25 June 1998.

21. Ibid.

22. Ibid.; and Paul interview.

23. McDermott, 25 June 1998; and Paul interview.

24. McDermott, 25 June 1998; and Paul interview.

25. McDermott, 25 June 1998; and Paul interview.

26. Aspin, 97–98.

27. David R. Stockman, director, OMB, to the heads of executive departments and establishments, letter, subject: Performance of Commercial Activities, 4 August 1983; and Executive Office of the President, Office of Management and Budget, Circular no. A-76, *Revised Supplemental Handbook: Performance of Commercial Activities,* March 1996.

28. Stockman; and *Revised Supplemental Handbook.*

29. Stockman; and *Revised Supplemental Handbook.*

30. Briefing, Maj Gen Richard R. Paul, director, Science and Technology, Headquarters Air Force Materiel Command, subject: S&T Manpower Reduction Strategy (FY 1997–2001), 16 February 1996; and Dr. Don Daniel, interviewed by author, 27 July 1998.

31. Briefing, Diane M. Disney, Deputy Assistant Secretary of Defense for Civilian Personnel Policy (DASD/CPP), subject: Civilian Workforce Issues: Briefing to the Defense Science Board's Human Resources Strategy Task Force, 21 December 1998.

32. Maj Gen Richard R. Paul, interviewed by author, 6 February and 2 March 1998.

33. Briefing, Col Mike Pepin, AFRL deputy director, Plans and Programs, subject: AFRL Manpower Report, FY 89–FY 01, 20 March 1998.

34. Bridgett Parsons, AFRL/Human Resources Office, interviewed by author, 27 July 1998; and Headquarters AFMC/XPMR, memorandum to Headquarters AFMC/XPMO, subject: Manpower Program Adjustment—Air Force Research Laboratory Phase I Authorization Transfer, with attached Air Mobility Command Manpower Program Subcommand Sequence, 30 May 1997.

35. Parsons interview.

36. Ibid.

37. Ibid.

38. Gen Henry Viccellio Jr., interviewed by author, 24 June 1998.

39. Paul interview, 2 March 1998.

40. Parsons interview.

41. Paul interviews, 6 February and 2 March 1998.

42. Ibid.

43. Ibid.; and Viccellio interview.

44. Pepin briefing.

45. Paul interview, 2 March 1998.

Chapter 5

Laboratory Studies and Strategy

Besides the inevitable reduction of civilians in the labs, another important issue facing the Air Force was organizational-management alternatives available to labs. Numerous studies on this subject had taken place over the years. Most recently, in 1993 Dr. George R. Abrahamson, chief scientist of the Air Force, led a study that had a particular bearing on the future of the four laboratories. Known as the Blue Ribbon Panel on Management Options for Air Force Laboratories, this investigation fulfilled a requirement levied on the Air Force by DDR&E to explore the feasibility of transforming the labs to a government-owned, contractor-operated management system. However, Abrahamson and the other eight distinguished panel members—three of whom had served as chief scientist of the Air Force—examined a broad range of management topics.[1]

Abrahamson's Blue-Ribbon Panel: Assessing Laboratory Management Structure

The blue-ribbon panel's findings strongly endorsed the value and contributions of Air Force labs. The members defined the role of the labs as providing critical leadership to ensure the development of the most advanced technologies and their integration into operational Air Force systems. They went on to explain that one of the "success factors" that contributed to the world-renowned reputation of the national laboratories—Los Alamos, Lawrence Livermore, and Sandia—was the deliberate design of a direct-reporting channel to a single point of contact at a high level. For example, Los Alamos reported to the president of the California Institute of Technology, and Lawrence Livermore had direct access to the provost of the Massachusetts Institute of Technology.[2]

On the other hand, the blue-ribbon panel faulted the out-of-step, multiple-reporting procedures followed by the Air Force labs. Rather than reporting to one powerful person in charge

Dr. George R. Abrahamson led a blue-ribbon panel study.

who had authority to make decisions in all areas, the labs divided their management responsibilities among several people at various levels of command. In the panel's opinion, these multiple-reporting lines weakened the overall operational effectiveness of Air Force labs: "Lab commanders report to the product center commanders for people, and to the TEO and SAF/AQ for program and budget. This leads to confusion and a lack of flexibility to manage the technical enterprize [*sic*], and greatly increases the difficulty of managing the drawdown in a rational way." According to the panel, splitting the key management aspects of the labs rather than reporting to a single person at the highest level of command fostered a perception in some quarters that Air Force senior leadership did not value its laboratories.[3]

Centralization of authority to control lab operations across-the-board was "urgently needed" to improve the overall operations of the lab system. One persuasive and strong-willed leader, whether a military commander or high-level civilian, armed with the flexibility to "move slots, redirect programs, and move funds without fighting organizational barriers" would be in a position to take action to cure a number of ills endemic to the labs. The panel bluntly stated that it wanted "a person for whom S&T stewardship is a full time responsibility." Doing that required prying the labs away from the product centers.[4]

Not everyone thought it was a good idea to separate the labs from the product centers. Gen Ronald W. Yates, AFMC commander, was not persuaded by the blue-ribbon panel's argument that laboratory operations would improve by having the labs report to the TEO (who ensured that labs operated as a single enterprise with a balanced investment strategy) instead of to the product centers. On 7 December 1993, Yates expressed his reservations after listening to Dr. Abrahamson brief the findings of the blue-ribbon study in the general's office. On 17 December, Abrahamson followed up with a letter to Yates "amplifying on the reasons" why the labs should operate as Air Force "corporate" assets as opposed to product-center assets.[5]

Abrahamson's letter did not change the general's mind. Yates wrote back, reiterating his "fundamental concerns" with the panel's recommendations. Because the general thought the labs were not broken, he saw no valid need for an internal reorganization: "With respect to changing the current linkage of the laboratories to the product centers, I remain convinced that, in the aggregate, the current reporting relationship is working well and should not be changed." In fact, Yates pointed out that the lab/product-center association "has unquestionably improved our track record in transitioning mature technology out of the laboratories to their customers." One of the things he especially liked about the current arrangement was the "frequent involvement of our product center commanders with their assigned laboratories," which accounted for the laboratories' high quality of work and responsiveness.[6]

Yates also pointed out that he relied on his director of S&T, who chaired the Science and Technology Mission Element Board, to work out all lab-related resource issues, including people and dollars. If people issues, such as personnel reductions, spilled over into the product centers, then one should use AFMC's Resource Allocation Integrated Product Team to resolve differing views. Failure to reach an agreement between the labs and centers would result in elevating the issue to General Yates for a final decision. Although the process was new, he wanted to give it a chance to work before changing the reporting relationship between the labs and centers.[7]

Yates did agree that the labs should serve a wide array of customers—not just their parent product centers. Accordingly, six months after writing to Abrahamson, General Yates issued a policy letter titled "Air Force Laboratories—Corporate Assets" but did not back off from the current lab-reporting arrangement. In the letter, he plainly stated that "our laboratories will continue to operate as organizational elements of our product centers." He firmly believed that this arrangement was the best way to "transition mature technologies to our weapon systems developers." After the general made his decision on this issue, no one could persuade him to change his mind. The policy remained in place until General Yates retired and departed the command on 30 June 1995. A year after General Viccellio took over as AFMC commander in July 1995, the pressures brought on by the National Defense Authorization Act and *Vision 21* reopened the issue of the labs' relationship to the product centers. It became increasingly clear that in the near future, labs would have to consolidate. As part of this restructuring process, labs would break away from their organizational alignment with product centers.[8]

New World Vistas: Building on *Toward New Horizons*

In November 1994, Gen Ronald R. Fogleman, chief of staff of the Air Force, and Sheila E. Widnall, secretary of the Air Force, sent a letter to Gene McCall, chairman of the Air Force Scientific Advisory Board (SAB). The underlying message of the letter was to remind McCall of the enormously significant role played by Gen Hap Arnold and Dr. Theodore von Kármán after World War II in establishing and promoting the importance of S&T in developing the Air Force of the future. *Toward New Horizons* was the vision von Kármán formulated to serve as the blueprint for conducting R&D programs that would lead to superior aerospace systems. The foresight embodied in *Toward New Horizons* laid the intellectual framework largely responsible for shaping the number-one Air Force in the world. Building on this legacy and recognizing the blistering pace of technological change in recent years, Fogleman and Widnall stressed that "only a constant inquisitive attitude toward science and a ceaseless and swift adaptation to new developments can maintain the security of this nation." Widnall strongly believed that the swift adaptation of technology preached by General Arnold was even more valid today than a half century ago. Widnall and Fogleman recognized this unswerving and time-tested principle of quickly applying technology as the underpinning of the global reach, global power policy of today's Air Force.[9]

Fogleman and Widnall sought to rekindle the spirit and attitude toward science that were so prominent during the von Kármán-Arnold era. Spirit, attitude, and preparation for the future were not empty concepts but important ideas of substance that the Air Force had to practice. If history served as any measure of truth, then Arnold's advice to the first meeting of the Scientific Advisory Group (forerunner of the SAB) in January 1945 held special significance. Appealing to the

Dr. Sheila E. Widnall, secretary of the Air Force, favored a major reorganization of the laboratories.

collective wisdom of the elite assembly of scientists, Arnold chose his words carefully and deliberately to make a major point in simple terms: "I don't think we dare muddle through the next 20 years the way we have the last 20." As a first step to applying Arnold's lessons learned to new possibilities in S&T, the chief and secretary challenged the SAB to devise a new vision for the next 30 years that focused on the most advanced air and space ideas which would transform the twenty-first century. This futuristic assessment of S&T would be detailed in a report called *New World Vistas,* scheduled for publication in December 1995, to coincide with and commemorate the 50th anniversary of the publication of von Kármán's *Toward New Horizons.*[10]

Essentially, Fogleman and Widnall charged McCall, study director of *New World Vistas,* and 150 eminent scientists from academia, industry, and government serving on the SAB team to develop their technology forecast in one year. The heart of the evaluation process focused on technology that most likely would "revolutionize the 21st century Air Force." Although the future of S&T remained the primary concern, Fogleman and Widnall also tasked the SAB to take a good look at how the laboratory system was organized to handle the management of

Dr. Gene McCall served as the study director of *New World Vistas*.

R&D programs. More specifically, they wanted the SAB to determine whether the laboratory structure was "consistent with the new vistas" findings. The fundamental question became, Should the Air Force make changes in the current laboratory organizational structure to better deal with projected technologies on the horizon?[11]

The answer to that crucial question was "yes." McCall and the SAB had no doubt that the demands of future technology would force changes in how the Air Force leadership approached structuring its organization to support R&D. Addressing the laboratory organization issue, the SAB offered constructive recommendations to best meet future contingencies. Like Abrahamson's blue-ribbon panel, the SAB wanted to remove the labs from the control of AFMC's product centers. Under the current organizational alignment, Rome Lab reported to the Electronic Systems Center, Phillips Lab to the Space and Missile Systems Center, Armstrong Lab to the Human Systems Center, and Wright Lab to the Aeronautical Systems Center. This organizational structure had its origins in 1982, when Air Force Systems Command placed laboratories under centers that reported to "product divisions." Later, with the establishment of four Air Force laboratories in 1990—and the elimination of centers such as the Air Force Space Technology Center, which managed the Weapons, Geophysics, and Astronautics labs—each lab now reported directly to a product center. By reducing the number of labs, the Air Force reasoned it could more effectively apply state-of-the-art S&T to better achieve mission success, streamline the acquisition process, reduce overhead, and eliminate duplication of technical efforts.[12]

Systems Command argued that it made a great deal of sense to place the labs under the product centers because doing so would force the labs to concentrate more on technologies that held the highest potential for transition to the operational Air Force. This setup had served its purpose well, according to the SAB. However, the board pointed out that each laboratory had "important programs which are not directly associated with its Product Center." Consequently, these types of programs would suffer because they were considered "outsiders." Looking to the future, the SAB reported that "the impact of new technologies is to demand closer integration and 'flattening' of organizations to provide better integration of the technologies themselves." Every senior Air Force leader, including Generals Viccellio and Paul, interpreted "closer integration and flattening of organizations" to mean a smaller and more centralized laboratory organization.[13]

From a philosophical and practical point of view, many people believed the time was right to shift labs away from the product centers. In the past, the Soviet bloc clearly represented the most dangerous threat facing the United States. During this time, it was realistic to assume that this near-term threat had the potential of erupting into a worldwide conflict. Because of this situation, emphasizing near-term S&T made sense. Under this arrangement, the labs would develop and pass on advanced technology as quickly as possible to the product centers to integrate with their systems and thus support the operational Air Force. The effort focused primarily on improving and evolving the next generation of existing war-fighting systems rather than developing revolutionary systems. This approach was analogous to car manufacturers turning out a new model every year to meet the near-term demands of customers, relegating long-term goals to secondary concern. General Motors and other car makers were not investing heavily in research to develop a radically new system, such as the electric car, to meet anticipated customer needs 30 years in the future. Instead, they sought to turn out new cars each year and get them into the hands of eager customers—the operating inventory—as quickly as possible.[14]

In the late 1980s, the political/military pendulum began to swing in a different direction. With the fall of the Berlin Wall in 1989 and the dissolution of the Soviet Union in 1991, the monolithic communist threat disappeared and was replaced by potentially more dangerous regional conflicts that the nation would have to deal with on a case-by-case basis. This change in strategy also affected how the Air Force thought about investing its S&T dollars during this new period of relative peace. Many believed the time was ripe to begin to dedicate an increased portion of the budget to promote higher-risk technology programs that would lead to revolutionary systems. The labs would still support the product divisions, but they did not need to be aligned with them organizationally. Centralized management of the labs, which seemed a more logical choice for achieving future short- and long-term goals, was the second major recommendation of the SAB.[15]

This recommendation involved placing the laboratories under an S&T executive who would have control over all laboratory personnel, funding, and programs. General Paul used this argument as one of the reasons to promote the single laboratory. As mentioned earlier, as the Air Force TEO, General Paul controlled funding and programs but had limited power over personnel decisions in the lab. Usually, when personnel cuts came down the line, the product-center commanders, rather than General Paul, wielded the final authority as to what positions would go or stay. This meant that, more often than not, the highest proportion of personnel cuts came out of the labs rather than the product centers. Paul and others believed that for the laboratory system to operate at the highest level of proficiency and productivity, the person directing the labs had to have total control over all funding, programs, and personnel issues. Paul consistently backed the SAB on this key issue, at every opportunity advocating that "it makes sense for one person to be accountable and work together in harmony with the manpower and budget aspects of the labs."[16]

The SAB believed that either a civilian or military person should hold this critical position. A civilian offered continuity over time, but a military executive could provide a closer connection to the operational Air Force. Although avoiding a commitment to either a military person or civilian, the SAB did take the stand that, wherever the executive came from, he should be at least the equivalent of a product-center commander. In managing programs, the S&T executive's most important responsibility was to better integrate technologies across-the-board and pressure his lab organization to pursue and increase transition opportunities. In other words, his most productive contributions would entail managing and developing those technologies that would lead to improved systems for supporting the war fighter—the bottom line for the existence of any Air Force laboratory organization. Equally important was the fact that the labs had an obligation to investigate and develop innovative technology to anticipate and meet the long-term demands of the Air Force over the next 30 years.[17]

Although everyone agreed that the labs should no longer remain a part of the product centers' organizational structure, the SAB made no specific recommendations suggesting how many labs should exist in the future. The board members did not believe they had the responsibility to tell the Air Force how to reorganize its laboratory structure. Consequently, the SAB issued no blueprint proposing a reduction of the existing four labs to three or two labs and made no official mention of establishing a single lab. Simply put, there was no opposition to either retaining four labs or establishing a single lab. Gene McCall remembered that the issue for reorganization emphasized the need to break the labs out from under the product centers and appoint an S&T executive who would provide centralized control over programs, people, and funding. The SAB wanted to flatten the organization and take steps to integrate similar technologies so the labs would become more responsive to the needs of the operational Air Force. The Air Force would decide how that happened.[18]

Providing better integration of technologies proved a somewhat elusive concept. In theory, most people agreed that in times of downsizing and restructuring the laboratory organization, it was a good idea to streamline operations by unifying technology efforts. As an example, both the Phillips and Wright labs conducted laser research. Portions of the work on sensors took place at the Wright, Rome, and Phillips labs. But the basic problem was that no one person, lab, division, or branch consolidated and directed all the various laser and sensors work taking place at the different geographic locations. Rather, both small and large groups of people worked independently, for the most part, on their portion of the laser or sensors pie with little or only minimal interaction with their colleagues at other sites.[19]

This complicated situation, which involved bits and pieces of the same technology discipline spread out among various labs, had been entrenched in the laboratory system for decades. Each lab built walls around the technology pieces it dealt with to prevent other labs from invading and taking away its prized work. Except to declare that this state of affairs needed to change, the *New World Vistas* study made no specific recommendations for breaking down the walls. However, McCall commented that restructuring the laboratory organization did not absolutely require the consolidation of similar technology work at one geographic site.[20]

Far more important to McCall in solving the problem was that scientists performing similar work at various locations make a much stronger effort to get to know one another and "work better together." To make that happen, one highly competent leader, appointed by the lab commander, had to assume responsibility for directing and managing all aspects of a particular technology discipline. That person would be the point of contact to answer any questions about specific technology posed by the Air Force, lab management, other government agencies, contractors, or any other customers. Of course, all this was easier said than done. Almost another year and a half would pass after the release of *New World Vistas* before General Paul and his staff agreed on a new single lab structured along the lines of specific technology directorates. Part of the reason for organizing in this fashion was to "better integrate technologies," as suggested by the *New World Vistas* study.[21]

Secretary Widnall remained extremely enthusiastic and supportive of the *New World Vistas* approach to S&T. Upon reviewing the study's findings, she definitely recognized the need to reorganize the lab structure so the Air Force would be better prepared to manage a diversity of emerging technologies over the next 30 years. She especially liked the idea of removing the labs from the product centers and making them more independent and productive. Faced with the prospects of declining budgets and fewer people in the workforce, she thought that the Air Force could achieve savings by consolidating all laboratory resources into one organization. Further, she believed that doing so would reduce the size of the administrative workforce and eliminate "non-value added administration" functions spread over all four laboratories.[22]

As Paul and Viccellio pressed to present their case in 1996 for a consolidated lab with a single person in charge, they won Widnall over without much convincing. Part of the reason had to do with her concerns about the technical duplication of effort among the various labs. She believed that one lab and one commander could correct this problem by making corporate decisions that would eliminate duplication of effort. At the same time, a single commander would have the authority and resources to promote an attitude of "greater technology sharing across Air Force missions" among scientists, who would respond better to a single lab's direction. Looking back, Widnall considered *New World Vistas* a very important and useful exercise that greatly influenced the decision to establish a consolidated laboratory in the Air Force.[23]

Global Engagement:
A New Strategy for the Next Century

Dr. Abrahamson's blue-ribbon panel and *New World Vistas* recognized that the Air Force laboratories would need to change their management structure and long-range vision to meet new challenges posed by political and military realities of the international arena. In response to the findings of these influential studies, the Air Force began taking steps to analyze its state of readiness for future combat contingencies. One of the most important actions taken by the Air Force was the decision to make a major shift in its long-range thinking by devising a new strategic plan called *Global Engagement: A Vision for the 21st Century Air Force.*

The two highest-level leaders in the Air Force, Secretary Widnall and General Fogleman, initiated a movement in May 1995 to produce a new Air Force vision for the twenty-first century. They assembled a study group consisting of senior military and civilian leaders and chaired by Gen Thomas S. Moorman Jr., Air Force vice chief of staff, to come up with a more realistic vision that would respond to changing political conditions around the world. Widnall, Fogleman, and others realized that they needed an updated strategic blueprint to better define the Air Force's role in DOD's "big picture" for fighting radically different future wars.[24]

To a large degree, *Joint Vision 2010,* issued by the Army's Gen John M. Shalikashvili, chairman of the Joint Chiefs of Staff, drove the urgency to develop a more modern Air Force vision. This publication provided overall direction for each service's responsibility in conducting future military operations. In other words, each service was an integral part of a joint team whose mission was to fight. Moreover, each service had to begin taking steps now to ensure that it would be prepared and able to fight to accomplish any mission assigned by the theater commander.[25]

Joint Vision 2010 appeared in the summer of 1996—the same time General Paul and his staff were preparing to present the single-lab concept to Corona—and demanded that the services use their collective capabilities to dominate the enemy in every aspect of the battlefield. This core concept, known as full-spectrum dominance, served as the foundation for planning and executing all future US military operations. Successful execution of this military doctrine depended on the four basic tenets of dominant maneuver, precision engagement, full-dimensional protection, and focused logistics—critical areas that each service had to strengthen by investing sufficient resources to become ready and capable of meeting any military contingency, anywhere in the world, in the first quarter of the twenty-first century.[26]

Since the end of the cold war, the Air Force had relied on its strategic policy of global reach, global power, which required execution of the four core competencies set out in *Toward the Future: Global Reach, Global Power:* air (eventually space) superiority, global mobility, precision employment, and information dominance. All of these had to be developed, sustained, and implemented when the call came. Secretary Widnall noted that this policy "has served us well" during a time when the Air Force maintained a high state of readiness, even though faced with significant downsizing of the force and reduced budgets over the first six years of the 1990s. But that policy could not go on forever. "Extraordinary developments in the post–Cold War era," Widnall stressed, "have made it essential that we design a new strategic vision for the Air Force." Rapidly changing geopolitical conditions worldwide dangerously affected the security environment and future missions of the Air Force. Perhaps most disturbing was the fact that the US military would not always be able to identify its enemy ahead of time, as it could during the cold war.[27]

Other challenges existed. Terrorist groups, whose actions were impossible to predict, clearly posed a major threat both at home and abroad. The proliferation and availability (for the right price) of advanced technology made it more likely that third world countries could acquire and use sophisticated weapon systems against a number of opponents, including the United States. As Secretary Widnall put it, "New technologies are erupting around us every day." Clearly, nuclear, biological, and chemical weapons were entering the arsenals of more and more nations around the world at an alarming rate. Additionally, DOD was gradually dismantling the once extensive forward-basing structure used by the United States and its allies; consequently, the United States would have to project military power from the homeland rather than strategic bases located around the world.[28]

After months of study and evaluation, General Moorman's vision team issued its new strategic blueprint—*Global Engagement*—in the fall of 1996. Widnall and Fogleman did not interpret that study as just another lofty statement of intent. Rather, this "action oriented" new strategic plan specified pathways for the Air Force to follow into the next century. *Global Engagement*, the culmination of the natural evolution of events, replaced *Global Reach, Global Power*, established in the summer of 1990.[29]

Six straightforward principles that the Air Force had to develop and put into practice over the next 30 years lay at the heart of the *Global Engagement* document. Four of the six core competencies that underlay the *Global Engagement* vision—air and space superiority, rapid global mobility, precision engagement, and information superiority—were modifications of the four *Global Reach, Global Power* core competencies. To make the strategic plan more robust, *Global Engagement* added

83

two other core competencies—global attack and agile combat support—to round out a total vision plan designed to deter aggression and to fight and win wars. In his explanation of the new vision to the Air Force Association in January 1997, General Moorman stated that "the context of the long-range plan is built around sustaining our *core competencies* and reinforcing the central themes found in the strategic vision."[30]

What effect did *Global Engagement* and the fundamental policy changes it brought about have on the Air Force laboratories? As it turned out, quite a bit. *Global Engagement* was one of many things that forced the laboratories to reevaluate how they could best support the six core competencies spelled out in the new strategic vision.

The Air Force leadership agreed that the six core competencies embodied in *Global Engagement*'s strategic vision correctly identified the service's goals and pathways. However, they represented only half the answer for solving future problems. To meet the future demands of airpower and space power, the Air Force knew it had to change the way it did business, which meant changing its culture and organization. Undoubtedly, change would come as DOD and Congress continued to pressure the Air Force to become more efficient with fewer resources and still successfully prepare for and fight future wars. Fewer resources translated to reduced budgets and fewer people—both military and civilian employees—assigned to all organizations across the Air Force, including the laboratories.[31]

Global Engagement was just one more example that drove home the point to Air Force leaders that the labs would have to reorganize. Over time, the Air Force had no choice other than committing to an aggressive reduction of infrastructure costs. Generals Viccellio and Paul certainly realized this fact of life and knew that they would no longer have the levels of money and people to support and sustain the current four-laboratory system. In terms of making the personnel workforce more efficient, the Air Force's position as stated in the new strategic vision called for changing the composition of the workforce. One change involved converting military positions in a variety of combat-support functions to civilian positions, thereby freeing military members to serve in the more critical operational jobs. This did not necessarily make for fewer positions on the unit-manning document, but contracting out to the private sector to perform combat-support jobs currently performed by military and civilians would have the effect of reducing the number of personnel—which would cut the number of positions on the unit-manning document. In addition, the Air Force believed that, in the long run, outsourcing and privatization of the civilian support positions could decrease a portion of overhead dollars. The intent of all these measures was to maximize efficiency by using the best-proven practices in the business world to run Air Force support functions. If the Air Force accepted the current business world's model of a modern corporation, then it would have to significantly reduce the size of its civilian support force.[32]

Secretary Widnall was very much aware and in favor of "getting rid of some administrative overhead structure. . . . That was an issue that was being looked at very carefully by the Office of the Secretary of Defense. . . . We were very well aware that in the view of many people, services [overhead], in general, had gotten bloated." Overhead was the abhorrent organizational fat that could not avoid drawing attention to itself. Overhead also ran counter to the military concept of a lean, mean, fighting machine. To illustrate the point, Widnall recalled, "You get this famous quote from one of those members of Congress from California who said there were more members of the acquisition force than the Marine Corps! When people say things like that, you know that has the potential to create the political pressure for major reduction in administrative overhead. There is no question about that." And there was no question in Widnall's mind that in view of the intense downsizing under way, the labs "needed basically to be looked at." The clear goal involved devising a better organizational strategy to keep the S&T arm of the Air Force in step with the vision and six core competencies to meet any demands on the battlefield during the next century.[33]

Too, projected budget and personnel reductions meant that the laboratories would have to operate as a leaner organization. Air Force Materiel Command simply could not afford to operate the labs as it did in the past because the reduced number of future dollars and people would not allow this to go on. Consequently, consolidation of laboratory assets into one central laboratory gained more and more favor at all levels of government. *Global Engagement* confirmed what Viccellio and Paul had already known and accepted. Looking to the out years, one could see that a single, consolidated lab made more sense in terms of economics and personnel than four separate laboratories operating independently; further, it would compete more and more for less money and fewer personnel over the next 30 years. Also, a centralized lab commander would have the authority to organize his agency by clearly defined scientific disciplines to help eliminate the overlap or duplication of technical efforts embedded for years in the four-laboratory setup. A single lab commander, with control over the total spectrum of S&T, would also be in a highly influential position to explore teaming alternatives with the other military services in an effort to increase efficiency by forming joint centers of excellence for R&D. These proposed centers of excellence would be established to directly support the core competencies articulated in *Global Engagement* and save money in the process.[34]

The goal of *Global Engagement* was not to destroy the laboratories—just the opposite. Perhaps the most important and revealing result emerging from *Global Engagement* was the declaration of change to the basic Air Force mission, which directly affected the laboratory missions. No longer was airpower exclusively the focus of attention. The new plan pointed to air *and* space superiority as the essential ingredient allowing all US forces freedom *from* attack and freedom *to* attack. The Air Force exists for the purpose of controlling what moves through air and space. As General Fogleman put it, "We have no other tasks. That is our only job. It is not a diversion for us. We do it full time—all the time." He was also extremely confident about S&T's future ability to adapt to new challenges and missions: "The reality is that in the first quarter of the twenty-first century it will become possible to find, fix, or track and target anything that moves on the surface of the earth."[35]

The starting point for sustaining air and space superiority unequivocally begins with the laboratory system that advances S&T through a diversity of strong R&D programs. Lab scientists and engineers serve as the eyes of the Air Force, looking into the future. These people draw upon their scientific expertise to build "combat capability" by coming up with inventive ways to improve existing systems, as well as introducing revolutionary systems to give the Air Force the technological edge to defeat any adversary. The absence of labs to keep the Air Force in the forefront of advancing and developing new technology for the war fighter would compromise success in combat. Today's modern arsenal of technology furnishes the military capability to achieve national objectives and, in the process, acts as the military silver bullet that saves lives.[36]

Secretary Widnall affirmed the implications of the Air Force change in mission. Knowing it was counterproductive to try to separate air and space operations, she realized that the focus had to remain on integrating air and space to maximize the unique assets of both to defeat the enemy. But for the future, she believed that the Air Force would find itself more dependent on space than air operations: "All this is just the beginning. We see, over time, our Air Force transitioning from an air force to an air and space force, and in the future, we fully expect to become a space and air force." With their technical know-how, the laboratories would lead the way by accelerating the development of innovative technologies to ensure the Air Force's capability to dominate space.[37]

Emerging from *Global Engagement,* the formation of six new battlelabs—small organizations designed to work closely with command customers to better define the operator's battlefield needs—helped with the development of innovative technologies to support air and space operations (fig. 4). General Moorman noted that each battlelab "is not a technology place" but a place where people with experience operating equipment and systems in the field generate new ideas. Focusing on "identifying innovative operational concepts that exploit mature technologies," the battlelabs work closely with the existing Air Force laboratory system to rapidly develop and test technical capabilities in the core-competency areas to demonstrate the pluses and minuses of the most promising operational concepts. Giving the war fighter in the field an opportunity to test a particular technology in its early development stage represented a new approach designed to eliminate all kinds of bureaucratic milestones and reviews in moving new technology through the start-up phases. If the war fighter tries it and likes it, then the lab can proceed with a more rigorous acquisition process. The idea is that this team approach will foster more realistic input from the user since the user of the equipment becomes an active participant in the process. The desired result entails taking the operator-level information, developing the technology, and modifying existing systems that will provide the person in the field with the most reliable and effective weapon and support systems. For example, the Space Battlelab at Falcon (now Schriever) AFB, Colorado, working with other laboratories, might be engaged in applying cutting-edge research from the labs on the geo-electro-optical deep-space surveillance system to keep better track of the number of satellites in orbit. Besides surveillance, other projects might evaluate innovative techniques for bolstering our space capability for the future, including weapons guidance, warning, communications, and environmental monitoring.[38]

According to Secretary Widnall, *Global Engagement* was a watershed event because it set a new course of action for the Air Force to follow. As part of this new vision, S&T would become an indispensable tool for shaping systems that the operational commands would use to fight wars in the next century. Moreover, it became absolutely essential that the laboratory infrastructure "seize these new technologies" that would "create combat capability from the metal and plastic of our equipment."[39]

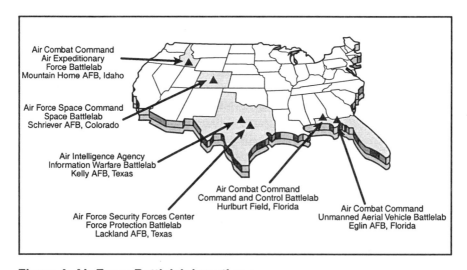

Figure 4. Air Force Battlelab Locations

Certainly, at the highest level of the Air Force, Widnall was out front proclaiming the virtues of S&T as one of the most important instruments for achieving victory in any future conflict. But she also realized—mainly because of significant budget cuts (estimated at about 40 percent), personnel reductions, and a much greater emphasis on the integration of air and space—that the laboratory system would have to change. A firm believer that "change is good," she favored Paul and Viccellio's reorganization plan "to put the research labs more directly into a single organization." She also thought that a single lab structure would be beneficial since it would stimulate more competition for funds—a healthy trend. In the past, the four labs knew that they would get their fair share of the budget. But by consolidating into one lab, budgets would probably go down, resulting in fewer lab dollars available. This was an effective way to ensure the highest quality work at the R&D level because only the most productive S&T programs would rise to the top for funding. The least productive programs would fall by the wayside, leaving the remaining limited funding that would go to support the more successful programs. Widnall liked this approach because "it is the only responsible way to manage the taxpayers' money to create basically internal competition for scarce resources."[40]

Secretary Widnall had no doubt that the time was ripe to pursue a fresh organizational approach to restructure the labs. Although she promoted the vision for the labs and the Air Force, her hectic schedule covering the entire spectrum of Air Force issues did not permit her to devote the time to work out the details to implement that vision—a job left primarily to General Paul. Starting in the summer of 1996, he began in earnest to muster his staff to build the rationale for establishing a single lab. This intensive effort would provide the basis of the proposal for a consolidated laboratory that would be presented at the Corona conference in the fall of 1996.[41]

Notes

1. Gen Ronald W. Yates, AFMC commander, asked Abrahamson to conduct the blue-ribbon study. Briefing, Dr. George R. Abrahamson, chief scientist, US Air Force, subject: Blue Ribbon Panel on Management Options for Air Force Laboratories, 10 January 1994.

2. Ibid.

3. Ibid.

4. Ibid.

5. Gen Ronald W. Yates, AFMC commander, to Dr. George Abrahamson, Air Force chief scientist, letter, subject: Blue Ribbon Panel, 4 January 1994; Dr. George R. Abrahamson to Col Richard W. Davis, commander, Phillips Laboratory, letter, subject: Blue Ribbon Study, 27 January 1994, with attachments (General Paul's charts, Mr. Jim Mattice's [deputy assistant secretary of the Air Force for research and engineering] chart, and Gen Merrill McPeak's speech). Yates served as AFMC commander from July 1992 through June 1995.

6. Yates letter.

7. Ibid.

8. Gen Ronald W. Yates, AFMC commander, policy letter, subject: Commander's Policy: Air Force Laboratories—Corporate Assets, 18 July 1994.

9. Gen Ronald R. Fogleman, Air Force chief of staff, and Sheila E. Widnall, secretary of the Air Force, to Dr. Gene McCall, chairman, Air Force Scientific Advisory Board, letter, subject: *New World Vistas* Challenge for Scientific Advisory Board, 29 November 1994; video, *New World Vistas*, prepared by Mr. Jim Slade and Maj Dik Daso, December 1995; and Dr. Sheila E. Widnall, "The Challenge," in *New World Vistas: Air and Space Power for the 21st Century*, ancillary volume (Washington, D.C.: Air Force Scientific Advisory Board, December 1995), 6–8.

10. Fogleman and Widnall letter; and Dr. Ivan A. Getting, "The SAB in Its Infancy," in *New World Vistas*, ancillary volume, 24–27.

11. The SAB consisted of 12 panels charged with evaluating future technologies: Aircraft and Propulsion, Attack, Directed Energy, Human Systems and Biotechnology, Information Applications, Information Technology, Materials, Mobility, Munitions, Sensors, Space Applications, and Space Technology. Fogleman and Widnall letter.

12. See Maj Dik Daso's interview with Dr. Gene H. McCall, chairman, Air Force Scientific Advisory Board, in *New World Vistas*, ancillary volume, 143–48.

13. *New World Vistas: Air and Space Power for the 21st Century*, summary volume (Washington, D.C.: Air Force Scientific Advisory Board, December 1995), 68.

14. Maj Gen Donald L. Lamberson, USAF, Retired, chair, Directed Energy Panel, *New World Vistas*, interviewed by author, 30 January 1998.

15. Ibid.

16. *New World Vistas*, summary volume, 68; Maj Gen Richard R. Paul, interviewed by author, 2 March 1998; and minutes of the Air Force Association Science and Technology Committee meeting, Crown Plaza Hotel, Dayton, Ohio, 17 July 1997.

17. *New World Vistas*, summary volume, 68.

18. McCall noted that a few SAB members with Air Force connections leaned toward supporting the current four-lab setup rather than establishing

a single lab. But the vast majority of SAB members did not advocate a specific reorganization plan. Dr. Gene H. McCall, study director, *New World Vistas,* interviewed by author, 13 May 1999; Lamberson interview; and Dr. Sheila E. Widnall, former secretary of the Air Force, interviewed by author, 7 July 1999.

19. Dr. Gene H. McCall, study director, *New World Vistas,* interviewed by author, 25 May 1999.

20. Ibid.

21. Ibid.

22. Dr. Sheila Widnall, Massachusetts Institute of Technology, to author, E-mail, subject: Single Lab, 28 June 1999; and Widnall interview.

23. Widnall E-mail; and Widnall interview.

24. Sheila E. Widnall, secretary of the Air Force, and Gen Ronald R. Fogleman, Air Force chief of staff, *Global Engagement: A Vision for the 21st Century Air Force* (Washington, D.C.: Department of the Air Force, November 1996), 1; and Sheila E. Widnall, "Adapting to an Altered Strategic Environment," speech delivered at the Center for Strategic and International Studies, Washington, D.C., 25 November 1996.

25. Air Force news release, "Air Force Secretary, Chief of Staff Unveil New Strategic Vision," 21 November 1996; on-line, Internet, 10 June 1999, available from http://www.af.mil/news/Nov1996/n19961122_961185. html; "Special Report: Global Engagement," *Airman Magazine,* February 1997; on-line, Internet, 7 June 1999, available from http://www.af.mil/lib/ globalon/index.html; and Gen John M. Shalikashvili, chairman of the Joint Chiefs of Staff, report, "Joint Vision 2010"; on-line, Internet, 18 June 1999, available from http://www.defenselink.mil/pubs.

26. *Toward the Future: Global Reach, Global Power,* US Air Force White Papers, 1989–1992 (Washington, D.C.: Department of the Air Force, 1993); Widnall and Fogleman, 5; and Gen Ronald R. Fogleman, Air Force chief of staff, "Global Engagement," speech delivered at the Smithsonian Institution, Washington, D.C., 21 November 1996.

27. Widnall and Fogleman, 1; "Air Force Secretary, Chief of Staff Unveil New Strategic Vision"; and "Special Report: Global Engagement."

28. Projecting firepower from the United States became a reality in the spring of 1999 with B-2s taking off from Whiteman AFB, Missouri, to perform bombing missions over Kosovo and then returning home after a 32-hour flight. Widnall and Fogleman, 1; and Widnall speech.

29. Gen Thomas S. Moorman Jr., Air Force vice chief of staff, "Implementing the Air Force's Strategic Vision," speech delivered to the Air Force Association Symposium, Orlando, Florida, 31 January 1997; and Gen Ronald R. Fogleman, Air Force chief of staff, "A Vision for the 21st Century Air Force," speech delivered at the Heritage Foundation, Washington, D.C., 13 December 1996.

30. Moorman speech.

31. Widnall interview; Widnall and Fogleman, 23; and "Special Report: Global Engagement."

32. Widnall and Fogleman, 19, 23.

33. Widnall interview.

34. Ibid.; Widnall and Fogleman, 23; and Widnall speech.

35. Widnall and Fogleman, 10; Fogleman speech, 21 November 1996; Fogleman speech, 13 December 1996; and "Special Report: Global Engagement."

36. Widnall and Fogleman, 10; Fogleman speech, 21 November 1996; Fogleman speech, 13 December 1996; and "Special Report: Global Engagement."

37. Widnall interview; and Widnall speech.

38. The other five battlelabs included Air Expeditionary Force Battlelab, Mountain Home AFB, Idaho; Command and Control Battlelab, Hurlburt Field, Fla.; Unmanned Aerial Vehicle Battlelab, Eglin AFB, Fla.; Information Warfare Battlelab, Kelly AFB, Tex.; and Force Protection Battlelab, Lackland AFB, Tex. "Battlelabs Created to Advance Air Force's Core Competencies," *Leading Edge,* February 1997, 11; and "Vice Chief of Staff Highlights Innovative Battlelabs," *Air Force News Service* (AFNS electronic bulletin board), 20 February 1997.

39. Widnall interview; and Sheila E. Widnall, secretary of the Air Force, "Beyond the Edge of the Map," speech delivered at the Smithsonian Institution, Washington, D.C., 21 November 1996.

40. Widnall interview; and Widnall speech, 21 November 1996. Although Widnall supported a consolidated lab for many reasons, she never felt pressured to form a single lab so the Air Force could compete at the same organizational level with one Army and one Navy lab. She did not discount the political importance of one Air Force lab but suspected that leaders of the Air Force S&T community worked this issue.

41. Widnall interview; and Widnall speech, 21 November 1996.

Chapter 6

Corona 1996:
Leadership and Decisions

To recap, by the summer of 1996, a series of events had transpired that, in one way or another, contributed to the Air Force's decision to move toward establishing a single laboratory. The most immediate circumstance affecting the single-lab decision was the passage of the National Defense Authorization Act of 1996, which subsequently led to the start of the *Vision 21* process in May 1996. However, prior to 1996, numerous independent studies and actions were percolating at different rates of speed throughout the Air Force's entire acquisition-management system. Starting in the mid-1980s, the Packard Commission recommended sweeping reforms in the acquisition-management process that resulted in creating an institutional mind-set which caused many high-ranking officials to reevaluate the role of laboratories. The Clinton administration also was determined to exert its influence since the president considered the laboratories a critical reform issue in his campaign to reduce big government. In 1993 he took steps to establish the National Science and Technology Council to devise new ways to inject more organizational efficiency into the management of laboratories. As it turned out, this was a slow and tedious process that did not lead quickly to tangible results.

Instead of making premature and radical changes to the lab structure and management principles, the Air Force turned to a number of studies conducted in the 1990s to assist in charting a realistic laboratory course for the future. These lab studies included George Abrahamson's blue-ribbon panel (1993–94), *New World Vistas* (1994–95), and *Global Engagement* (1995–96). Although each study put its own particular spin on lab management, all of them agreed that laboratories definitely would have to restructure their management system to meet and survive the demands placed on them in the twenty-first century.

In the meantime, while studies and debates wore on, trying to map out the best course for the future of S&T, the labs struggled under the weight of a number of mandates imposed by the Office of the Secretary of Defense to reduce personnel. Implementation of the Dorn cuts, the A-76 process, and the defense planning guidance steadily chipped away at the labs' personnel infrastructure. This went on for several years, but the lab leadership knew that the existing system could not withstand this constant assault, which weakened the way the labs conducted their day-to-day business. Midway through 1996, the laboratory system reached a crossroads.

General Viccellio, convinced that the timing was right with the appearance of *Vision 21*, set into motion two important actions designed to radically change the lab-management system. He did this first by presenting a new laboratory organizational setup to the Air Force's top leaders gathered at Corona 1996, held at the Air Force Academy in October. A month after Corona, he laid out the specifics of his plan to reform the labs to the secretary of the Air Force. This second presentation of his lab vision was a response to the *Vision 21* requirement that asked how the Air Force planned to consolidate its labs for the future. Viccellio knew that if his lab plan were to succeed, he had to rely primarily on General Paul and his staff to provide the substance and rationale for drastically altering the lab-management structure. Paul and his staff immediately set to work, spending the entire summer assessing various lab options. The outcome of all this was a persuasive strategy that proposed forming a single lab, accepted by the secretary and chief of staff in November. This intensive summer of work laid the foundation for the eventual stand-up of the Air Force Research Laboratory.

Before consolidation received approval, however, a great deal of work and planning had to take place at the command level. Because of *Vision 21*, General Viccellio had to give the Office of the Secretary of Defense the specifics of the Air Force's plan for restructuring its labs by November 1996. General Paul certainly realized the importance and implications of *Vision 21*: "There was a stake in the sand that said we, the Air Force, had to go back to the Office of the Secretary of Defense and tell them what we would do within the Air Force with things under our control—not cross-service control." With this clear tasking at hand, General Paul and his office became the operations center for developing a lab master plan that Viccellio could brief to any audience. To get this vitally important process under way, Paul selected Tim Dues in June to head a team to determine what the ideal laboratory structure should look like during the first quarter of the twenty-first century. Each of the four laboratory commanders appointed representatives to serve on this "white paper" study group chaired by Dues. Besides the chairman, participants included Christine Anderson, Norm Sorenson, Pat Nutz, Glenn Harsberger, Ken Boff, Chuck Helwig, B. F. Gould, and Igor Plonisch.[1]

In July, this study group met at Wright-Patterson AFB to brainstorm and discuss ideas for incorporation into the white paper. Lieutenant Colonel Nutz, representing Wright Laboratory and its XP shop, stressed that, over the past year, he had led a Wright Lab team that had visited and listened to 14 executives explain how they ran their companies. Nutz sought to gain better insight into how these successful managers in the private sector structured their workforce to attain maximum organizational efficiency.[2]

Nutz found a deliberate move by companies over the past few years to consolidate their resources internally and buy out competitors when the opportunity presented itself. In 1993, for instance, Martin Marietta acquired GE Aerospace. Shortly thereafter, one of the largest mergers of all time occurred when Lockheed and Martin Marietta combined to form Lockheed Martin on 15 March 1995. A year later, Lockheed Martin acquired Loral. Boeing, the world's largest aircraft company, followed a similar consolidation pattern, purchasing North American Rockwell in 1996. Two years later, Boeing made another major reorganizational move by purchasing McDonnell Douglas. Consolidating assets then led to the downsizing of overhead, especially in the middle-management ranks. This reorganization pattern was a fact of life that businesses had put into practice in order to scale back and survive in a highly competitive environment with scarce funds. After assessing all the information he had collected from the private sector, Nutz strongly advocated that the Air Force laboratory system take a page from the playbook of industry by pursuing a policy of consolidation. This fundamental change in thinking, according to Nutz, would lead to improved organizational efficiency at the laboratory level.[3]

Most members of the study group shared Nutz's reasoning that consolidation was the course the laboratories should set for the future. However, *consolidation* remained a somewhat elusive term not yet completely defined. The fact that it did not necessarily equate to a single lab stimulated plenty of discussion covering a wide range of options for laboratory reorganization. Possibilities included moving to two labs (one each for air and space), three labs, or one lab. Another alternative suggested leaving the four labs in place and reducing the number of personnel assigned to each. Some even flirted with the idea of doing away with the entire Air Force laboratory system and combining elements from all service labs into one "purple" lab centrally controlled by the secretary of defense.[4]

The study group's white paper concluded that the best option for the Air Force entailed creating a single laboratory to centralize technology unique to military applications. A single lab would require less overhead than would several laboratories, and in the long-run, it would cost less. The Air Force would gain organizational efficiency, the study group argued, by operating the single lab more along the lines of a large private corporation. As a start, a board of directors in Washington, D.C., would run the lab and establish the appropriate political connections. The board's president would have responsibility for all funding, personnel, and programs and would head a small long-range planning group that would work closely with the Office of the Secretary of Defense. Other board members would serve as "directors" responsible for a specific technology program.[5]

The white-paper team also insisted that the labs divorce themselves from the four existing product centers, which systematically avoided high-risk technology programs to solve short-term technology problems. On the other hand, a laboratory by definition engaged in high-risk, high-payoff R&D to push the militarily relevant technology edge. A single lab with a military commander or civilian director (each similar to a company's board president) would have the authority to reduce manpower and budgets as well as get rid of noncore technologies in quick fashion. The lab could achieve other savings by directing more investments to modeling and simulation research, which equated to nonphysical means of research—and thus reduce the physical infrastructure (buildings, equipment, test devices, etc.) it needed to conduct research. Also, a single-lab commander or director would be in a much better position to coordinate and strengthen lab ties with both industry and academia.[6]

The participants who put the white paper together presented arguments for a single lab in very broad terms. Driven by *Vision 21*, the white paper represented only a first step to start people thinking about the makeup of the current lab structure and what needed changing. As noted in the white paper's appendix, no one had sufficient time to address all the details of a new lab organization. Essentially, the white paper was one of the first attempts by General Paul's staff to systematically put together a collection of ideas for building an Air Force response to *Vision 21*'s tasking to consolidate laboratories. The heart of that response—as proposed in the white paper—was the Air Force's need to establish a single lab. General Viccellio quickly agreed. As he explained, he and General Paul had discussed the possibility of a single lab as soon as *Vision 21* had come out. They had repeatedly weighed a number of options for the future of S&T but kept coming back to the single lab as the best solution for reducing personnel, centralizing control with one commander, and producing quality technology.[7]

Another very important event paralleled the development of the white-paper project. General Fogleman had identified 17 major topics distributed among five panels for discussion at the fall Corona meeting. He appointed General Moorman, the vice chief, to chair a long-range-planning board of directors responsible for assuring preparation of all the issue papers in the same format for presentation at Corona. Moorman selected Lt Gen Lawrence P. Farrell Jr., Air Force Materiel Command's vice commander, to head panel three to develop AFMC's Corona papers nine through 12 on key acquisition and infrastructure issues facing the Air Force in the future. Corona Issue Paper 9 specifically addressed laboratory reorganization. These papers would serve as the basis for generating discussions among the

Gen Ronald R. Fogleman, chief of staff of the Air Force, backed the lab-reform movement.

attendees at the Corona conference, scheduled to take place at the Air Force Academy 8–12 October 1996.[8]

Vision 21 and Corona were separate but interrelated events. They were separate in the sense that Corona was strictly an internal Air Force forum to give top commanders an opportunity to brief the chief of staff and secretary of the Air Force on topics that would affect current and future missions. Corona addressed a number of complex questions, including the matter of deciding how the Air Force intended to streamline its entire acquisition process. Management of laboratories represented only one subset of this larger acquisition question open for discussion at Corona. *Vision 21,* however, was a top-down tasking from Congress and DOD that required AFMC to focus exclusively on what the Air Force planned to do to consolidate its laboratory system. The service had to complete and brief that plan to the secretary of the Air Force in November 1996. So, both Corona and AFMC's response to *Vision 21* tried to solve the lab-consolidation issue, but consolidating the labs was only a small part of many Corona meeting topics that needed resolution, while developing a response to *Vision 21* was the sole issue that AFMC needed to resolve.[9]

Corona Issue Paper 9, "Weapon Systems Acquisition, Science & Technology & Associated Infrastructure," was just one of many topics on the agenda scheduled for discussion at the upcoming meeting at the Air Force Academy in October. Maj Gen Robert E. Linhard, special assistant to the chief of staff of the Air Force for long-range planning, had the responsibility for coordinating with all the commands to provide 17 issue papers that answered very specific questions covering a range of Air Force concerns. As mentioned earlier, to make this process more manageable, General Moorman had set up five panels: Core Competencies/Air, Space/Information Warfare, Acquisition and Infrastructure, People Values and Career, and Organization and Force Mix. Panel three, Acquisition and Infrastructure, focused on reengineering the acquisition and sustainment processes to meet required capabilities efficiency. This panel gave General Viccellio an excellent opportunity to present his views on restructuring the laboratories to the most influential Air Force leaders.[10]

As chairman of panel three, General Farrell received assistance from Lt Gen George K. Muellner, principal deputy, Office of the Assistant Secretary of the Air Force (Acquisition); Robert F. Hale, assistant secretary of the Air Force for financial management and comptroller; Maj Gen Eugene A. Lupia, civil engineer, Headquarters US Air Force; and Brig Gen Daniel P. Leaf, deputy assistant chief of staff, US Forces Korea. Maj Gen Richard N. Roellig, who headed AFMC's procurement shop, acted as General Farrell's principal action officer for getting Issue Paper 9 in its final format, which consisted of seven sections: issue statement, scope, desired/potential objectives, key factors affecting decision, decision options, summary of options analyzed (pros and cons, etc.), and relationship to other issues.[11]

One of the main objectives the Air Force set forth to reform its acquisition system was creation of a leaner S&T infrastructure that would result in flattening the existing organizational hierarchy to the greatest extent possible. This had special implications for the labs because they were an integral part of the acquisition process. General Paul knew that, to a large degree, the future of the labs depended on the issues presented and decided at Corona. With that in mind, he again assembled his most trusted and experienced advisors—Tim Dues, Dr. Daniel, Dr. Russo, and Colonel Markisello—to lead the exercise to make sure that realistic options on the labs' future found their way into Issue Paper 9.[12]

They considered numerous lab-consolidation options that covered the entire spectrum of possibilities—from having the Air Force turn over partial control of the labs to contractors to getting out of the laboratory business completely. Under this move toward privatization, three options appeared. The first entailed establishing a government-owned, contractor-assisted agreement whereby the contractor would manage all lab-support activities. Also, the contractor would assume a larger role in supporting technical programs and would gradually make up a larger percentage of the contractor/government workforce ratio. However, the government would continue to make all decisions on the management of the laboratory. A second option involved a government-owned, contractor-operated infrastructure. Under this scheme, contractors would take over as sole technical program managers and would operate all government facilities. For the government to retain overall control, the contractor would have to report directly to the government's consolidated weapon-systems management staff. The third option entailed divesting almost complete control of the labs to the contractor, who, in this case, would assume responsibility for the total system program, including development and delivery of required weapon products. In effect, the government would become a customer with only minimal influence over the contractor.[13]

Another set of consolidation options rejected any movement toward contractor control of labs and endorsed a plan whereby the Air Force would continue to retain ownership of the weapons-management process but would reorganize to form a more centralized system for managing its laboratories. One proposal called for the establishment of three laboratory centers of excellence to manage weapon systems from "cradle to grave" (concept phase through retirement). This scheme would include space; air; and command, control, communications, computers, and intelligence (C⁴I) centers and would assign every weapon system to one of these centers. This plan also envisioned that all organizations assigned to support a particular center, such as a system program office, would geographically consolidate at a single location to support its center for the entire life cycle of the weapon systems. A second consolidation option—a modification of the first—called for establishing two lab centers of excellence: one for space and one for air systems. This too would be a cradle-to-grave operation, as would the third option—triservice management of the laboratory centers of excellence. One service (Army, Navy, or Air Force) would act as the lead executive responsible for a joint

100

center of excellence comprising air vehicles/engines/human systems/avionics/applicable technology areas; satellites/spacelift/missiles/directed-energy weapons/applicable technology areas; C⁴I; electronic warfare; and munitions. Personnel from all three services would work at each laboratory center, but the lead executive service would own the infrastructure (buildings, people, money, and programs). Still another option had DOD creating one laboratory, managed by the deputy director for defense research and engineering, that would accommodate all service needs.[14]

For a number of reasons, none of these options appealed to General Viccellio. Regardless of whether contractors or the government ran the labs, he and others believed that an awkward and cumbersome management structure would evolve. On the positive side, some options could produce gains in terms of reduced staffs and overhead, less duplication, and better cooperation among the three services on S&T matters. But the negatives seemed to predominate. Allowing contractors to take over the labs involved huge political hurdles. That is, increases in the contractors' authority would likely lead to more layoffs of government workers, which, in turn, would instigate political opposition. In addition, putting more weapon-systems management authority in the hands of contractors would tend to weaken the Air Force's link to the war fighter. In short, unlike contractors, the military had the best interest of the war fighter at heart.[15]

Setting up two or three lab centers of excellence would also be politically painful—not to mention expensive—because moving people to one of the centers would set into motion RIF actions and base-closure procedures. Furthermore, many people worried that creating a triservice lab center of excellence— thereby becoming partners with the other two services, who would naturally promote their agendas—would weaken the Air Force's institutional interests. In other words, the Air Force ethos would be lost. Finally, moving control of S&T up the chain of command to the DOD level would not be prudent because it would further remove military users from the weapon-systems management staff.[16]

Viccellio and Paul were convinced that no good reason existed for turning the laboratory system over to contractors or fixing the problem by moving lab management up the chain of command or outside the Air Force. Both believed that the three lab options they put on the table were more realistic. Tab 9-12 of Corona Issue Paper 9 clearly spelled out the three options addressing laboratory structures and reporting arrangements. Early drafts of the issue paper incorporated the first two options, while the third did not appear until the end of July.[17]

All three options aimed to create an infrastructure that capitalized on applying best practices to ensure best value for the Air Force. The first option favored keeping multiple labs in place under the current system. Each of the four labs would maintain alignment with its acquisition center of excellence: Wright Lab with the Aeronautical Systems Center, Phillips Lab with the Space and Missile Systems Center, and so forth. There would be no change in the lab commanders reporting to the TEO—in this case, General Paul—on all issues dealing with the management of S&T programs. Likewise, lab commanders would continue to report to the center commanders, who would attend to the care and feeding of the laboratory workforce, which included personnel and facility matters. Overall, the first option amounted to keeping the current laboratory organization in place. But maintaining the status quo did not mean that the labs would continue to conduct business as usual. Although the same lab framework would remain, everyone realized the inevitability of a significant decline in the current lab workforce over the next few years.[18]

Good reasons existed for keeping the current lab structure in place. One could implement consolidation of support and overhead functions with relative ease. By having the labs report to and work closely with existing centers, one could argue that the Air Force was on the right track by advancing near-term technology that could turn into useful products for use by customers—the operating commands—now rather than 20 years in the future. Near-term weapon systems appealed to military commanders and congressmen, who seemed more willing to allocate funding that would lead to improved systems capable of providing better protection of our troops in the field. The current lab system also came in for praise because of the mutually beneficial relationship it had fostered with industry and academia over the past 20 years. Air Force–sponsored science programs at private companies and universities had resulted in the return of advances in technology to the Air Force to enhance system development, which, in turn, led to improved products.[19]

Despite the obvious advantages of keeping the current lab system intact, one could not ignore some major drawbacks—particularly the fact that organizational seams among labs would not go away. Work on the same groups of technologies (lasers, electronics, signal processing, materials, etc.—identified as crosscutting technologies) occurred in multiple labs. People often perceived this fragmented approach as a duplication of effort and resources. Moreover, it made it difficult for General Paul as TEO to plan and control technology efforts spread across several labs since no one person could tell him about all aspects and progress of one particular technology discipline.[20]

Further, lab commanders found themselves in the awkward position of working for two bosses—the TEO and the center commander—which also meant a splitting of S&T resources. The TEO controlled dollars and program management, but people and facilities fell under the purview of the center commander. "This situation," as the Corona paper pointed out, "complicates and prolongs people/position/facility/dollar/allocation decisions leading to suboptimized decision making." Viccellio and Paul wanted to correct this divided control of lab resources. They believed that splitting responsibilities violated the goal of "best practices," which the Air Force strove to achieve as part of its overall acquisition-reform movement.[21]

Besides the drawback of working for two bosses and splitting management responsibilities, under the multiple lab setup, each lab employed its own support staff. The plans organizations, located at each lab, consisted of a decentralized setup that, according to many people, should be consolidated into one centralized office. In a time of diminishing resources, the clear trend in government called for reduction of excessive overhead and consolidation of like functions. Keeping four laboratory staffs appeared excessive, inconsistent, and out of step with the basic lab-reform policy that the president, Congress, and DOD advocated across the Air Force. In addition, option one ignored recommendations from two important studies of the future of the laboratories: Abrahamson's blue-ribbon panel and the SAB's *New World Vistas* (both discussed earlier). Neither study favored having lab commanders report to two bosses; rather, each commander should report directly to a single S&T executive. Although option one did not endorse the single-reporting concept, it found it viable because that approach would cause the least disruption to the existing laboratory system.[22]

Option two, the next step up in the evolutionary plan of lab consolidation, did incorporate the advice of the blue-ribbon panel and the SAB by supporting the idea that each lab commander would report to a single S&T director who would serve as both the TEO and the S&T executive at Headquarters AFMC/ST. Thus, the TEO remained in charge of both programs *and* people. Positions on the center's manpower listing would become part of the laboratory's unit-manning document—a big change because now the S&T executive (General Paul) not only would control programs but also would "own" the people (scientists, engineers, and support personnel) as well as all the labs' facilities. However, option two retained a portion of option one—alignment of each lab with its acquisition center of excellence. Thus, one could expect a minimum of organizational changes under option two because multiple laboratories would remain in place.[23]

Giving all resources—people, dollars, and facilities—to a single executive-in-charge would result in a centrally controlled operation that could provide much better integrated planning and decision making across the entire S&T organization. Option two emphasized that "more integration across Divisions, Directorates, and Laboratories is needed" as well as a "better capacity for addressing multidisciplinary problems." This would not happen, the argument went, as long as labs reported to a number of center commanders. A single S&T person could alter the decentralized, integrated planning approach followed by center commanders who pursued their own R&D agenda—primarily focused on meeting short-term technology goals.[24]

Given the right amount of authority and accountability, a single S&T leader would be in a position to objectively balance all S&T resources to support users in the near term. That leader could also set direction for the long term to identify and sustain technology's "push requirements" that would lead to revolutionary technical systems. Furthermore, this arrangement would enhance "the S&T Executive's influence by raising his visibility and that of S&T to the same level as the product, test and logistics centers." Although all this seemed a move in the right direction, a single S&T executive would not solve all the laboratory problems under option two, as pointed out by the Corona paper's summary: "Multiple labs, even with a single S&T Executive, provide a suboptimal organizational structure for optimizing [Air Force] crosscutting technologies."[25]

Option three became part of Corona Issue Paper 9 at the end of July. As *Vision 21* and Corona progressed simultaneously, General Paul wrote General Viccellio, suggesting that it would be beneficial to add the single-lab consolidation option to the Corona paper. To help build the Corona issue papers, General Farrell had asked for inputs from all offices that had a stake in the labs' future. However, the first draft did not include the single-lab option. After reviewing the initial draft of the issue paper, General Paul noticed that it made no reference to the single lab. He wanted to correct that omission because he saw Corona as the perfect opportunity to test the waters on the single-lab proposal. In his letter to Viccellio, Paul urged that the single-lab plan be one of the main options included in the issue paper:

> Specifically, we could create a single Air Force laboratory (which subsumes our existing 4-lab structure) reporting to a single laboratory commander who reported to AFMC/CC or the Service Acquisition Executive. The Army has a single lab: Army Research Lab; the Navy has a single lab: Naval Research Laboratory. Thus the Air Force analogy would be the "Air Force Research Laboratory (AFRL)." . . . A single laboratory would not only show parallelism with the other two services, but would also maximize the synergy of "corporate" technologies that support both air and space, e.g. materials, electronics, photonics, geophysics, C^4I, human systems and manufacturing. . . . I simply offer it [the single-lab option] for completeness, and because periodically the question has been asked "Why doesn't the Air Force have a single lab like the other two Services?"[26]

General Paul wanted the single-lab option included in the Corona papers for two main reasons. Firstly, he deeply believed that establishing a single lab was the best option for restructuring the laboratory system in terms of reducing personnel and improving organizational efficiency. But Paul also viewed the Corona process and meeting as a golden opportunity and a timely mechanism to give the single-lab concept exposure in front of the four-star leadership, as well as the Air Force chief of staff and the secretary. Paul and Viccellio certainly realized the advantage of raising the single-lab concept at Corona because of its tie-in with *Vision 21*. Essentially, Corona became a dress rehearsal for finalizing the *Vision 21* briefing that General Viccellio was scheduled to have ready for the secretary in November. With the single lab briefed at Corona, Widnall and Fogleman for the first time would have a chance to think about the single-lab option. Their reaction at Corona to this radical proposal would be critical. If they agreed with the concept, General Viccellio would know that he was on the right course for restructuring the lab organization. Introducing the single-lab option at Corona also would preclude any surprises for the secretary when she received a briefing in November on the Air Force's internal strategy for *Vision 21*.[27]

Option one advocated keeping multiple labs, and option two endorsed a similar plan; but with a single-lab commander, the most dramatic change in laboratory organization appeared in option three. Unlike the others, it proposed establishing one corporate Air Force laboratory headed by a single commander. Generals Viccellio and Paul considered option three "the most consolidated lab configuration" that would best meet the needs of an Air Force facing declining budgets and personnel resources. This plan called for merging the existing four laboratories and the Air Force Office of Scientific Research into one laboratory. Organizationally, "major directorates of like technologies" would replace the four labs. The new single-lab commander would be elevated to an organizational level equivalent to that of product-center commanders and would report directly to the AFMC commander. No longer would AFMC's Science and Technology Directorate be a staff agency but would convert to the "command section" of the single lab. Finally, the laboratory commander would control all resources (program funding, people, technology programs, and facilities) in order to achieve maximum flexibility and consistency of decision making in the day-to-day running of the new organization.[28]

One of the most potent arguments for a single lab was that it would "eliminate organizational seams (and duplication) between cross-cutting technologies now worked in multiple labs." For example, various aspects of advancing signal-processing technology took place simultaneously at different labs to support different missions. Phillips Lab engaged in signal-processing work to support satellites; Rome Lab used it to improve C⁴I systems; and Wright Lab depended on it to develop better aircraft avionics. Setting up a single lab with technology directorates authorized to manage similar technologies—previously scattered among many labs—would help eliminate these technology seams. As stated in the analysis section of the Corona paper, getting rid of the seams would "result in an optimum use of limited funds by eliminating duplication and the cost/delays in trying to pass technology across the artificial seams." In other words, a single technology director in the lab who managed all aspects of a particular technology across-the-board would be in the best position to make decisions affecting program priorities and funding of various components of one particular technology discipline. However, the downside of forming a number of technology directorates, which would surely outnumber the four labs, was that the single-lab commander would have an increased span of control in managing all the technology directorates. As it turned out, after formation of the single lab, the commander had to deal with 10 technology directors instead of four laboratory commanders.[29]

Option three used the same reasoning as option two in promoting the value and advantages of putting a single commander in control. Detached from the control of the product-center commander, a single-lab commander would have the authority to make decisions on all S&T resources—programs, funding, people, and facilities. This arrangement positioned the lab commander to capitalize on the best management practices, which in turn would result in ensuring "best value" in the procurement of advanced-technology weapon systems. In addition, a single-lab commander would speak as one voice for the organization in establishing timely responses to "the needs of the marketplace, [major commands], industry and academia, and make [Air Force] labs more competitive with the other services to obtain work supportive of [Air Force] objectives." By clarifying and strengthening the lines of communications, the single lab would stand to gain a better reputation in the academic and industrial communities. This would allow the commander to exert greater leverage to draw upon these two labor pools of scientific knowledge to benefit the Air Force. On the negative side, industry and academia might view a large single lab as a monolith that would prove difficult to approach. But the counterargument held that the lab commander could cut through the bureaucratic red tape by directing industry and academia to a specific technology directorate, thereby removing the perception of the lab as an unwieldy government bureaucratic institution that took a long time to make decisions.[30]

Making decisions to establish better relationships and support with private corporations and universities was only one part of the communications picture. A new lab, removed from the oversight of the product centers, could "better support *all* AFMC organizations" (emphasis in original). The single-lab commander would be able to "objectively balance" programs, funds, and personnel, thus supporting users at all levels. Transition of technology from the lab to all the product centers would continue to meet their real-time requirements for integrating it into new systems. But at the same time, the single lab would move off in new directions, becoming more actively engaged in investing heavily in technology programs designed to produce midterm and long-term breakthroughs to revolutionize the development of futuristic weapon systems. By following this approach, one Air Force lab would operate similarly to the Army Research Lab and the Naval Research Lab. The Air Force feared, however, that DOD would see its lab as an inviting target to take over and merge with the Army and Navy labs, creating one DOD lab. Such an assimilation into what, in effect, would be a single triservice laboratory would weaken the Air Force ethos.[31]

Formation of a DOD lab might have loomed on the distant horizon as a major long-term problem, but no immediate problem existed in terms of increasing funding to form a single Air Force lab. Since forming a consolidated lab constituted an "organizational realignment," the service would incur no additional costs to implement option three. Minor additional costs might arise in the physical consolidation of some planned technology directorates, but savings gained from reduced overhead costs during the consolidation would most likely offset them.[32]

In many ways, presenting a convincing case for a single laboratory at the upcoming Corona conference was very similar to a trial lawyer's making his closing argument before a jury. In this case, General Viccellio had to make a convincing summation of his position to the Air Force jury consisting of Secretary Widnall and the undersecretaries who accompanied her; General Fogleman; and the nine commanders of the major commands. By September, Viccellio began shaping his own unequivocally clear and persuasive closing argument advocating the value of creating a single lab:

> One corporate Air Force laboratory with a single commander would serve the needs of multiple [chief executive officers] and all S&T customers. It satisfies all of the objectives and factors identified as necessary for the acquisition community. Primarily, it brings single ownership of all S&T related resources (i.e., funds, people, and facilities) to the single lab commander for fully integrated planning. It eliminates organizational seams between cross-cutting technologies now worked in multiple labs while reducing overhead and improving efficiency by eliminating separate lab command sections and planning staffs and the TEO planning staff. Further, it raises the visibility and stature of S&T within AFMC to the level of the other centers supporting all AFMC organizations. The negatives are the potential for de-emphasizing technology transition to the SPOs [system program offices] residing in the lab's current parent product center (although strong corporate technology transition processes have been institutionalized by the TEO for all labs over the past five years) and DMR; that is, the commander as TEO is accountable for programmatics *and* training, organizing and equipping responsibilities. This option is the ultimate lab consolidation possible while maintaining [Air Force] ethos.[33]

Although General Viccellio had clearly articulated his position on the single lab, he did not brief this issue at the Corona meeting in Colorado Springs. Since laboratory restructuring— along with other acquisition and infrastructure issues—fell under panel three, General Moorman directed General Farrell, as chairman of that panel, to present the panel's briefings. Farrell's area of responsibility included briefing four issue papers covering a complexity of Air Force acquisition and sustainment processes: (1) processes and systems that will best support power

Lt Gen Lawrence P. Farrell Jr. briefed laboratory options at Corona.

projection in 2025; (2) future basing structure; (3) core technology and evaluation capabilities (infrastructure) needed to continue acquisition of superior weapon systems; and (4) management of the acquisition infrastructure at all levels, including reorganization of the laboratory system. A few weeks prior to Corona, each of the four-stars received a stack of notebooks containing all the issue papers that would be discussed at the meeting. The idea was that all the generals would study the key issues ahead of time and be prepared to comment on topics with which they had the most concern.[34]

Farrell and Viccellio, the only two people representing Air Force Materiel Command at Corona, sat together at the same table. When it came his turn to brief, Farrell spent the next four hours going over the four issue papers assigned to panel three. Most of this time focused on presenting a diversity of acquisition options that had nothing to do with laboratory restructuring. Farrell recalled that he spent only about three minutes briefing how the laboratory system should be changed to better meet Air Force missions of the future, but those three minutes represented the first time the single-lab concept was officially presented to the highest-level assembly of Air Force decision makers.[35]

At Corona, the heart of the argument for consolidating all Air Force laboratories into one lab appeared on a single slide that succinctly summarized General Viccellio's reasons for proposing a single lab (table 4). One of the most important benefits of creating one laboratory was that it would streamline the current lab structure by reducing overhead. Moreover, appointing a single commander to lead the new organization meant that this person could exert more effective leadership and control over all lab resources: people, dollars, programs, and facilities. Finally, under this more centralized management

Table 4

End State No. 3:
Single Air Force Laboratory

- Combines AFOSR and four Air Force labs into a single Air Force lab

- Provides a streamlined structure—reduced management overhead
  — Single commander
  — Single staff

- Consolidates full resource ownership and accountability

- Reduces fragmentation of similar technologies

---

approach, the commander could reduce fragmentation of simi-lar technologies spread out among several geographic sites.[36]

Putting up the single-lab slide in full view of all the Corona attendees was the first real test in determining how the high command would react to the radical proposal of doing away with four labs and combining them into one. Whether or not the attendees supported the new lab, their input would prove very influential in determining the future organization of the laboratory system. Naturally, General Viccellio was intently interested in the group's collective response—particularly Secre-tary Widnall's and General Fogleman's reaction to the single-lab proposal.[37]

The anticipated lengthy debate on the single-lab proposal at Corona never materialized. Farrell and Viccellio were prepared to vigorously defend the laboratory consolidation plan, but no one opposed the proposal laid out on the single slide. As Farrell described the scene, the proposal to establish one labo-ratory "got very little discussion," and everyone in the room agreed that "it sounds like a pretty good idea." "Everyone" included Widnall and Fogleman, who liked the idea of consoli-dating diverse lab resources into a single lab. Farrell recalled that "they gave us a head nod on the spot." Viccellio didn't have to make a big speech to sell the single lab—it sold itself. According to Farrell, Viccellio "could see the water flowing in that direction [one lab], so he just let it flow." Viccellio had no doubt that General Fogleman and Secretary Widnall certainly supported the single lab, noting that "everyone thought that it was the right thing to do because it [consolidation] was similar in concept to so much else that was going on in the Air Force."[38]

Actually, the makeup of the Corona audience served to limit the discussion of the single lab. The majority of general officers there were interested primarily in war-fighting doctrine that affected the operation of their commands. They were more comfortable dealing with operational aspects such as missile defense, communications, space, airpower tactics and strategy, maintenance, information management, and similar issues. The operational commanders acknowledged the importance of science, technology, and the complicated acquisition process, but they simply were not conversant with all of the details and perplexities. To them, acquisition appeared one step removed from the operations side of the Air Force. Secretary Widnall certainly understood the position of the operational commanders and the reason for the lack of discussion of or objection to the single lab proposal: "You wouldn't expect it. You don't expect the PACAF [Pacific Air Forces] Commander to object to this [one lab]. On what basis is the PACAF Commander going to object to the Air Force Materiel Commander proposal to consolidate the Air Force labs, which in fact report directly to him? That's not going to happen!"[39]

General Viccellio was very pleased with the outcome of the Corona briefing on the single lab. He had accomplished what he had set out to do by presenting the single-lab proposal to the most influential leaders in the Air Force and getting their initial reaction to the plan. Although the response was positive, Viccellio knew this was only the first step toward creating one lab. Secretary Widnall interpreted what went on at Corona not as an official approval of the single lab but as a "general consensus. Nobody saw any serious flaws in what was being proposed."[40]

General Viccellio now needed General Paul and his staff to develop a comprehensive plan providing the details of how four laboratories would transition to one. That laboratory plan, which would also form the basis for the Air Force's response to *Vision 21* on laboratory reform, was scheduled for presentation to Secretary Widnall in November. Everyone realized that this would be an extremely important meeting. The secretary would have to make a decision, either allowing the Air Force to move forward with the creation of a single lab or rejecting the proposal and directing the service to proceed on a different course. However, because of the favorable reaction to the single lab at Corona, most people were convinced that Widnall would have no choice other than to officially sanction the new lab in November.[41]

Notes

1. Maj Gen Richard R. Paul, interviewed by author, 2 March 1998; Tim Dues, interviewed by author, 6 April 1998; Lt Col Pat Nutz, interviewed by author, 4 February 1998; and "The Air Force Laboratories in 2025," white paper [August 1996].

2. Nutz interview.

3. Ibid.; George Williams, Lockheed-Martin (Albuquerque office), telephone conversation with author, 12 August 1999; Paul Toth, Boeing Aircraft (Albuquerque office), telephone conversation with author, 12 August 1999; and "Aviation Industry Evolution," briefing chart provided by Paul Toth, 16 August 1999.

4. Nutz interview; and Maj Gen Richard R. Paul, interviewed by author, 6 February 1998.

5. Dues interview; and "The Air Force Laboratories in 2025."

6. Dues interview; and "The Air Force Laboratories in 2025."

7. Gen Henry Viccellio Jr., interviewed by author, 24 June 1998.

8. Ibid.; Lt Gen Lawrence P. Farrell Jr., USAF, Retired, interviewed by author, 24 August 1999; and Gen Ronald R. Fogleman, USAF, Retired, interviewed by author, 14 January 2000.

9. Paul interview, 6 February 1998; and Fogleman interview.

10. Maj Gen Robert E. Linhard, special assistant to the chief of staff of the Air Force (CSAF) for long-range planning, to board of directors (BOD), letter, subject: Corona Issue Paper Update, 11 July 1996; "CSAF CORONA Decision Papers," Air Force internal working paper [June 1996]; and briefing, General Moorman, Air Force vice chief of staff, subject: Air Force Long-Range Planning Corona Top Update to BOD, 20 June 1996.

11. Moorman briefing; and Paul interview, 2 March 1998.

12. Tim Dues, interviewed by author, 12 March 1998.

13. Ibid.; and Corona Issue Paper 9, "Weapon Systems Acquisition, Science & Technology & Associated Infrastructure," draft, Air Force internal working paper, 26 July 1996.

14. Dues interview, 12 March 1998; Corona Issue Paper 9; and Col Dennis F. Markisello, interviewed by author, 6 February 1998.

15. Corona Issue Paper 9.

16. Ibid.

17. Ibid.

18. Tab 9-12 of Corona Issue Paper 9, "Weapon Systems Acquisition, Science & Technology & Associated Infrastructure," draft, Air Force internal working paper, 28 August 1996.

19. Ibid.

20. Ibid.

21. Ibid.; Viccellio interview; and Paul interview, 2 March 1998.

22. Tab 9-12 of Corona Issue Paper 9, 28 August 1996.

23. Ibid.

24. Ibid.

25. Ibid.

26. Maj Gen Richard R. Paul to AFMC (General Viccellio), letter, subject: Another Laboratory Option for the AF/LR Initiative, 29 July 1996.

27. Paul interview, 6 February 1998.

28. Tab 9-12 of Corona Issue Paper 9, 28 August 1996.

29. Ibid.

30. Ibid.

31. Ibid.

32. Ibid.

33. Ibid.

34. Farrell interview; and "CSAF CORONA Decision Papers," Air Force internal working paper, July 1996.

35. Farrell interview.

36. Ibid.; briefing slide, "Single [Air Force] Laboratory," presented by General Farrell at Corona Conference, 8–12 October 1996; and Viccellio interview.

37. Viccellio interview.

38. Ibid.; Farrell interview; Dr. Sheila E. Widnall, interviewed by author, 7 July 1999; and Fogleman interview.

39. Widnall interview; Farrell interview; and Markisello interview.

40. Widnall interview; and Viccellio interview.

41. Viccellio interview; and Paul interviews, 6 February and 2 March 1998.

Chapter 7

The Last Dance:
Meeting in the Secretary's Office

No sooner had the Corona conference ended than Generals Viccellio and Paul turned their attention to preparing for the meeting with Secretary Widnall on 20 November, when they hoped to secure her approval to move forward with the single-lab option. Paul and his staff had only four weeks or so to finalize their *Vision 21* working paper that would provide valuable input in designing the briefing for the secretary.

In the summer of 1996, General Paul and his staff were in the process of conceptualizing and developing the initial draft of a plan that would lay out how the Air Force intended to respond to the *Vision 21* tasking to consolidate into as few labs as practicable. Vince Russo—who led General Paul's *Vision 21* team—and his group had the responsibility for putting the working paper together. However, development of the *Vision 21* plan did not occur in isolation. Numerous coordination meetings took place in order to share ideas with the other *Vision 21* points of contact throughout the Air Force. Alan Goldstayn, deputy director of Plans and Programs at AFMC, served as General Viccellio's *Vision 21* action officer. At the secretary of the Air Force level, Arthur Money, assistant secretary of the Air Force for acquisition, appointed Blaise Durante, deputy assistant secretary for management policy and program integration, as the Air Force lead on *Vision 21*. Durante relied on Lt Col Walt Fred as his primary assistant for working *Vision 21* taskings.[1]

Although a great deal of interaction occurred at all levels of the Air Force with regard to *Vision 21*, General Paul and Vince Russo's team played a pivotal role in preparing the initial working paper, which went through several revisions. During October, the paper carried the title "The Air Force Theodore von Kármán Laboratory: A Strawman Overview," but by November it had changed to "The Air Force Laboratory: An Overview." However, the basic content of all versions of the working paper remained the same in terms of subjects addressed, the first of which was the organization of the four existing laboratories, followed by the proposed single-lab organizational structure headed by one lab commander and consisting of 10 technology directorates: Air Force Office of Scientific Research; Human Health; Space; Command, Control, Communications, Computers, Intelligence, Surveillance, and Reconnaissance (C⁴ISR); Materials/Processes; Electronics; Optics; Flight Dynamics; Propulsion and Power; and Weapons (fig. 5). A comparison of the existing lab structure to that of the new single laboratory showed the Air Force's seriousness about radically changing the way it would do business in the future.[2]

The core of the *Vision 21* working paper focused on a plan for setting up the new technology directorates in the single laboratory. Defining a realistic vision for the entire lab represented one of the most important responsibilities of the lab commander, who also had the job of managing technical programs, dollars, people, and facilities, as well as ensuring that the headquarters

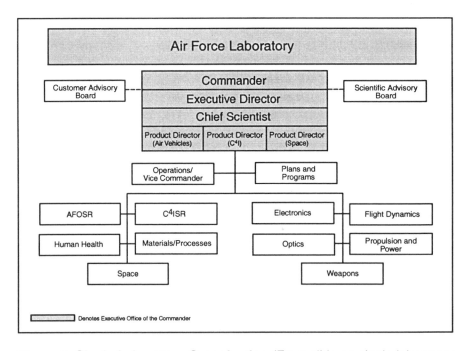

Figure 5. Single-Laboratory Organization (From slide on single laboratory organization, "The Air Force Laboratory: An Overview," Air Force *Vision 21* working paper, [November 1996])

staff functioned effectively. Under the commander, 10 technology directorates would replace 26 directorates, the "model" technology directorate consisting of anywhere from three hundred to seven hundred people and supported by two to five technical divisions (one hundred to three hundred people) as well as an integration and operations division. A Senior Executive Service (SES) employee would lead each tech directorate, assisted by a full colonel as deputy director. A similar personnel arrangement held true for the technical divisions, each of which would have two or more branches (25 to 50 people), depending upon its mission. General Paul and other high-level Air Force officials believed that during the initial stages of consolidating into one laboratory, the Air Force could achieve a savings of up to seven hundred personnel positions. After the new lab had operated for several years, projections put its total workforce at five thousand people.[3]

Although details about the single lab began taking shape in the *Vision 21* working paper, two key issues still needed resolving. The first concerned the future of AFOSR: should it function as a separate directorate, or should its current resources be distributed among the technology directorates? One argument maintained that the new lab's chief scientist should oversee the basic research program currently managed by AFOSR. Each technology directorate could effectively manage contracts to universities for providing basic research support to the laboratory. On the one hand, dismantling AFOSR as a separate organization would likely result in manpower reductions. On the other hand, breaking it up would mean that AFOSR most likely would lose its strong presence and influence in Washington, D.C.'s basic research community.[4]

Dr. Joe Janni, director of AFOSR, appealed to General Paul that it made more sense for his office to function as a directorate within the new laboratory. Janni pointed out that each technology directorate under the single lab would have to reduce its staff as part of the laboratory-consolidation process and that each directorate would have only a small staff organization to execute its mission. Because he already had a "very flat" organization due to downsizing at the end of 1995, AFOSR seemed a prime candidate to fit the definition of the model technology directorate envisioned in the single lab.

Later, during the implementation phase of the lab, Janni's position prevailed, and AFOSR became the equivalent of a technology directorate.[5]

A similar fate occurred with munitions technology. There was some talk of integrating munitions work (performed at Eglin AFB, Florida, and controlled by Wright Laboratory) into the appropriate technology directorates of the new lab. The advantage of this approach was that it would streamline operations by combining like technologies for sensors, guidance, structures, and so forth, as well as producing manpower savings. The downside of spreading munitions technologies among the various technology directorates was that it would weaken their optimal integration. As with AFOSR, Weapons (later Munitions) did not become one of the 10 projected technology directorates that would make up the Air Force Research Laboratory until the spring of 1997.[6]

The future of AFOSR and Weapons remained undecided in the final version of the *Vision 21* working paper released in November. However, General Paul was convinced that the new lab model presented in that paper met the intent of what had transpired at Corona in October. His next step called for assisting in preparing and finalizing the briefing for Secretary Widnall on 20 November in order to obtain her approval of the single-lab proposal. But the job of briefing the secretary fell to Blaise Durante rather than General Paul. In putting the briefing together, Durante relied heavily on input from Paul and information contained in the *Vision 21* working paper. As Durante explained it, the briefing sought to answer the question, What did the Air Force plan to do to reduce, restructure, and revitalize an apparently bloated R&D infrastructure?[7]

With only a few weeks available to assemble the briefing, Durante's approach entailed hitting the "high points, not the minutiae" in trying to convey to the secretary the prudence and timeliness of endorsing the idea of combining four laboratories into one. As Durante bluntly put it, "When you are at the secretary level, you have 30 minutes—get your point across, give it to the secretary, and get the hell out!" After all, Secretary Widnall had already heard the single-lab pitch at Corona and offered no opposition to the plan at that time. Consequently, Durante and others anticipated that the briefing would be just a formality they had to go through before proceeding with the implementation of the consolidated laboratory.[8]

Even though expectations were high that the secretary would approve the new lab, Durante still put a lot of work into the briefing to make it as persuasive as possible. A few weeks before the briefing, Paul met with Durante in his office in the Pentagon to review the *Vision 21* working paper and put together the briefing slides. Durante knew that General Paul was the most knowledgeable person on all aspects of the single lab—both the big ideas and details—and wanted to tap his expertise to sort out the most cogent points to present to Secretary Widnall.[9]

The slides that Durante and Paul developed went right to the point. The first couple of charts covered background on the National Defense Authorization Act of 1996, *Vision 21*, and the president's directive to develop a plan and schedule for downsizing DOD's labs. One of the fundamental points that Durante had to underscore as the centerpiece of his presentation was that as a result of *Vision 21*, DOD had directed the Air Force to consolidate its labs into "as few installations as is practicable and possible." This process had to start now and reach completion by 1 October 2005. Durante also intended to remind the secretary that laboratory downsizing was already under way with the 35 percent R&D personnel reduction imposed by the defense planning guidance. Looking to the future, a 20 percent reduction (DOD's first *Vision 21* estimate to reduce positions) with the right mixture of personnel would have significant consequences for laboratory operations. It would mean that customer support in technology products and services would jump from 45 percent of the laboratory's infrastructure to 56 percent, which placed more emphasis on S&T. By reducing infrastructure 20 percent, lab overhead would drop from 45 percent to 36 percent of the lab's total operation (fig. 6).[10]

Durante also made full use of historical perspective to drive some key points home. For instance, he borrowed a slide that General Paul had often used to show that, because the labs had taken significant manpower reductions since 1989, institutional reform had been under way for years. The lab workforce of 8,493 dropped to sixty-three hundred in 1996, and

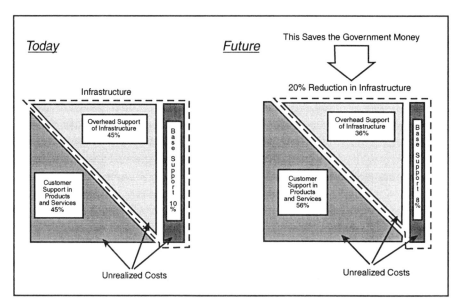

Figure 6. *Vision 21* **Background Infrastructure** (From briefing, Blaise Durante to Secretary Widnall and General Fogleman, subject: Air Force Strategy Meeting, 20 November 1996)

projections indicated it would reach 5,507 by 2001. Another historical chart clearly illustrated how the laboratory system had evolved over the years (fig. 7). Faced with growing pressure from Congress, the Clinton administration, and DOD, the Air Force thought that going to one lab seemed consistent with the natural evolution of the laboratory system.[11]

Once Durante was satisfied with the initial set of briefing slides, he wanted to do a test run of the briefing to obtain feedback from top Air Force leaders. A few days before the meeting on 20 November, he briefed General Moorman, the Air Force vice chief of staff, to get his reaction to the single-lab proposal. Durante went through the entire briefing, emphasizing that the Air Force had to commit to reducing, restructuring, and revitalizing its R&D infrastructure functions, activities, and facilities to meet the Air Force's vision and missions of the future. The most effective way to accomplish this, Durante pointed out, was to reorganize and consolidate resources by establishing a single laboratory. Such a facility, controlled by one commander, would be in a better position to bring

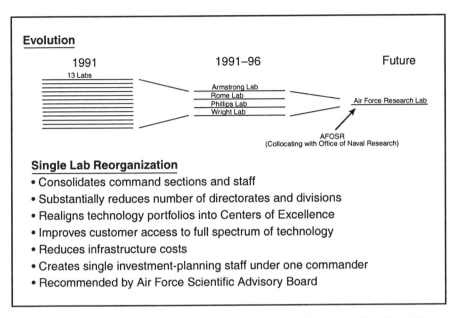

Figure 7. S&T Strategy—Single-Laboratory Reorganization (From briefing, Blaise Durante to Secretary Widnall and General Fogleman, subject: Air Force Strategy Meeting, 20 November 1996)

about meaningful change. The commander would have to pursue a policy of divestiture whereby the lab leadership would continually evaluate technology thrusts and programs and eliminate those that offered little potential for leading to new systems for directly supporting the war fighter. Further, the lab would have to establish an aggressive outsourcing program to draw upon the technical expertise of private contractors to make the lab as productive as possible. Finally, in a drive to increase efficiency, the lab would have to tailor its organization to apply better business practices that had been tested and proven in the private sector.[12]

General Moorman's reaction was predictable. Listening to Durante and fully aware of what had transpired at Corona, he endorsed the briefing and did not believe the secretary would have a problem with it. He knew that the Air Force was under substantial pressure to do something about the labs. After months of assessing various options, a single lab seemed a workable and timely solution to the tasking imposed by *Vision*

Blaise Durante briefed Secretary Widnall on the single lab on 20 November 1996.

21. The notion of centralizing control of the labs through a single commander seemed especially appealing. In short, Moorman did not anticipate any opposition from the chief or secretary and thought the single lab was a done deal.[13]

On Wednesday morning, 20 November 1996, General Paul and others gathered in Secretary Widnall's office in the Pentagon to hear Durante's briefing on the single-lab proposal. The atmosphere was cordial and all business as the small but influential group took their seats. Mr. Money and General Muellner were present to answer any questions that Secretary Widnall or General Fogleman might ask on what effect a consolidated laboratory might have on the acquisition process. The secretary's aide—Colonel Fred—and a few other administrators were also present.[14]

Showing the slides that he and General Paul had prepared, Durante hit the high points, stressing to the secretary that the Air Force had to come up with a solid position in response to the requirements of *Vision 21.* Durante's main point was that creating a single lab would improve the management of Air Force R&D across-the-board and at the same time reduce the number of people in the lab workforce by as many as seven hundred during the first phase of reorganization. In addition, the single lab would gain more prominence by being elevated to the same level of authority as product centers, test centers, air logistics centers, and specialty centers (fig. 8). The briefing went smoothly and was over in less than 45 minutes. Widnall had a question on what reaction "the Hill" might have to the formation of a consolidated Air Force laboratory. General Paul assured her that the likelihood of members of Congress opposing a new lab was remote, especially since the Army and Navy each had a single lab. In short, there was very little discussion

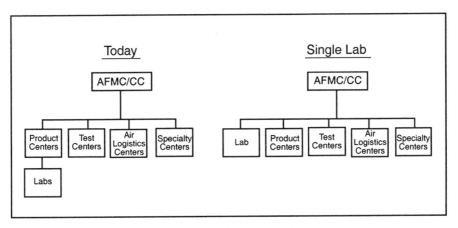

Figure 8. Old versus New Laboratory Structure: AFMC Organization
(From briefing, Blaise Durante to Secretary Widnall and General Fogleman, subject: Air Force Strategy Meeting, 20 November 1996)

on any of the key issues in Durante's presentation, signifying everyone's satisfaction with the proposed single-lab concept.[15]

This meeting represented a major turning point in terms of how Air Force laboratories would conduct business in the future. General Paul remarked that they undoubtedly had Secretary Widnall's official approval of the single laboratory: "The Secretary did ask a couple of clarifying questions at the 20 November meeting. It was apparent to me from her head nods and verbal acknowledgments during the briefing that she was very supportive. She did approve the single lab concept." Durante confirmed Paul's recollections, remembering the secretary's nodding her head in agreement and stating that the single lab concept "looked good" and was "a workable plan." She then informed Durante that the single lab was the official Air Force position in response to *Vision 21* and told him he had her approval to brief Dr. Anita Jones, director of defense research and engineering at DOD, on how the Air Force intended to restructure its laboratory system.[16]

Five days after the meeting with Secretary Widnall, Durante did just that. Since this briefing was at DOD level, he wanted to make sure he tied the consolidated lab proposal to the larger issues of vision and mission (fig. 9). Thus, one of the main points he emphasized was that the Air Force mission,

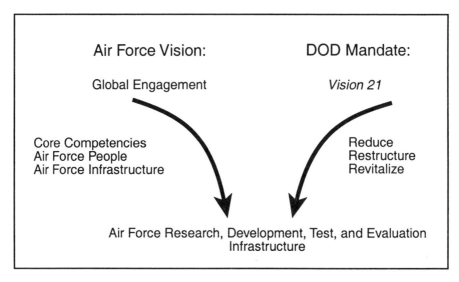

Figure 9. Merging Vision and Mission (From briefing, Blaise Durante to Dr. Anita Jones, director of defense research and engineering, subject: Air Force Strategy: *Vision 21*, 25 November 1996)

driven by *Global Engagement* and its core competencies, fitted extremely well with DOD's *Vision 21* mandate directing each of the military services to reorganize its laboratory system. A single lab, Durante explained, would be consistent with the newly evolved strategy of *Global Engagement* and at the same time would meet the consolidation demands of *Vision 21*.[17]

Dr. Jones expressed some skepticism about the consolidated laboratory plan, especially the Air Force's optimistic projection of a workforce reduction of seven hundred people soon after the laboratory stood up. Durante sensed that Jones and her staff had doubts that the Air Force would really take steps to remove that many positions from the unit-manning document. She thought that the Air Force would eventually end up fixing the books rather than getting rid of seven hundred people who filled real jobs. In short, she found herself on the horns of a dilemma. On the one hand, because she wanted the lab's personnel numbers to come down to comply with *Vision 21*'s strategy, she should enthusiastically support the single-lab concept. On the other hand, she knew that large personnel reductions could produce political repercussions, as had

BRAC 95. Thus, she compromised, giving a "tentative OK" to proceed with the single lab. Durante interpreted her decision as DOD's approval for the Air Force to move forward with its intraservice plan—its response to *Vision 21*—to create a consolidated laboratory. Jones had the option at any time of reviewing the Air Force's lab-implementation plan and recommending that it be stopped. However, over the next few months, as the idea of the single lab began to unfold and show more merit, she chose not to interfere with the plan.[18]

After the meeting with Secretary Widnall, General Paul felt a great deal of relief and accomplishment. Months of hard work by Paul, his key staff members, and others had come to a successful conclusion with the secretary's decision to move forward with the single lab. Elated with Secretary Widnall's approval, Paul returned to Wright-Patterson and immediately wrote a letter to his staff and four laboratory commanders, informing them that the secretary and chief had "approved organizational consolidation of the 4 labs and AFOSR into a single laboratory as part of the Air Force's *Vision 21* strategy." General Viccellio sent out a similar letter on 26 November, announcing that the secretary and chief had "approved the single laboratory concept as an element of the Air Force's *Vision 21* strategy." The top-level goals of the new single lab, Viccellio pointed out, were to streamline the laboratory structure by reducing overhead, decrease the fragmentation of similar technology work at multiple geographical locations, and put dollars and people under the control of a single commander.[19]

Viccellio also identified single-lab tenets for use as essential guidelines in setting up the new lab. These included eliminating the four existing labs and their command sections, appointing a single-lab commander, replacing the four labs' plans organizations with a single plans office, reorganizing the current 25 technology directorates into 10–12 large directorates, and moving all S&T personnel to a new single-lab manpower document.[20]

General Viccellio counted on his director of S&T to implement the secretary's decision. Paul did not waste any time laying the groundwork for the formation of the single lab, insisting that it be set up as soon as possible. Accordingly, he directed his staff and lab commanders to meet with him "to initiate the detailed planning process" for setting up the new laboratory. Scheduling the meeting for 5 and 6 December at the Bergamo Center, several miles from Wright-Patterson, Paul told the invitees of the importance and "great opportunity ahead" for making significant changes to the S&T infrastructure. Looking to the future, he envisioned that dismantling the old organization and creating a new one would be a "corporate" S&T effort. Since the secretary had made her decision, no one should debate the wisdom of moving to a single lab—there was no turning back. Everyone had an obligation to support the decision and utilize his or her talents to make the new lab happen. No one person could do all this alone. It would take a steady team effort, with groups at all organizational levels pulling together to effect the transformation. Paul, who knew this better than anyone else, was anxious to get started because he and his staff had to identify, discuss, and resolve a variety of issues before they could put together a workable lab-implementation plan.[21]

Notes

1. Blaise Durante, deputy assistant secretary of the Air Force for management policy and program integration, Office of the Assistant Secretary of the Air Force for Acquisition, interviewed by author, 3 September 1999; and Lt Col Walt Fred, Office of the Assistant Secretary of the Air Force for Acquisition, interviewed by author, 3 September 1999.

2. "The Air Force Theodore von Kármán Laboratory: A Strawman Overview," Air Force *Vision 21* working paper, 30 October 1996.

3. "The Air Force Laboratory: An Overview," Air Force *Vision 21* working paper, [November 1996]; and Durante interview.

4. "The Air Force Laboratory: An Overview."

5. Joseph F. Janni to Headquarters AFMC/ST (Maj Gen Richard R. Paul), letter, subject: Laboratory Consolidation, 9 December 1996.

6. "The Air Force Laboratory: An Overview."

7. Durante interview; and Fred interview.

8. Durante interview.

9. Ibid.; and *Vision 21*, Air Force internal working paper, 8 November 1996.

10. Durante interview; and *Vision 21*, Air Force internal working paper, 8 November 1996.

11. Durante interview; and *Vision 21*, Air Force internal working paper, 8 November 1996.

12. Durante interview; and *Vision 21*, Air Force internal working paper, 8 November 1996.

13. Durante interview.

14. Ibid.; Fred interview; and Dr. Sheila E. Widnall, interviewed by author, 7 July 1999.

15. Durante interview; Fred interview; Widnall interview; Gen Richard R. Paul, interviewed by author, 2 March 1998; and briefing, Blaise Durante to Secretary Widnall and General Fogleman, subject: Air Force Strategy Meeting, 20 November 1996.

16. Maj Gen Richard R. Paul to author, E-mail, subject: More Answers, 30 March 1998; Durante interview; and Gen Ronald R. Fogleman, interviewed by author, 14 January 2000.

17. Briefing, Blaise Durante to Dr. Anita Jones, director of defense research and engineering, subject: Air Force Strategy: *Vision 21*, 25 November 1996.

18. Durante interview; and Fred interview.

19. Maj Gen Richard R. Paul, director of Science and Technology, to [see distribution], letter, subject: Single Lab Planning Offsite, 22 November 1996; and Gen Henry Viccellio Jr. to [see distribution], letter, subject: Single Air Force Laboratory, 26 November 1996.

20. Viccellio letter.

21. Paul interview; and Paul letter.

Chapter 8

Conclusion

The decision to create the Air Force Research Laboratory resulted from a series of events and judgments that began in the mid-1980s and ended in November 1996 with the secretary of the Air Force's approval of the single-lab proposal. No one dramatic event was responsible for the formation of the AFRL. Instead, over the years a number of studies, legislative matters, reports to the president, and DOD directives—all of which penetrated various levels of government—in one way or another affected the eventual birthing of the single laboratory.

Changing the S&T culture in the Air Force was a slow and tedious process. One can trace the beginnings of laboratory reform to the Packard Commission's report, released 10 years prior to the formation of the AFRL. Packard's message was clear: to keep pace with better business practices used by the private sector, DOD's antiquated acquisition system needed to undergo substantial changes in the way it conducted business to become a more cost-effective and productive organization. In an effort to reform the acquisition system, Congress passed the Goldwater-Nichols Department of Defense Reorganization Act in October 1986, requiring DOD to comply with the Packard Commission's recommendations. In February 1989, Congress passed the Gramm-Rudman-Hollings Act, which gave President Bush authority to direct the secretary of defense to devise a strategy to make sweeping reforms in DOD. The secretary completed a major reorganization plan, known as the Defense Management Review, that committed to making across-the-board changes, including implementation of the Packard findings. DMRD 922 (October 1989) advised the Pentagon to give serious consideration to merging all military labs directly under DOD—all in the name of good economics. Projections of declining budgets over the next decade demanded fundamental changes in order to improve organizational efficiency by reducing the number of assigned personnel.

One of the first indicators of meaningful laboratory reform occurred on 13 December 1990, when 13 Air Force laboratories

129

merged into four. Less than two years later, in July 1992, another major reorganization took place when Air Force Logistics Command combined with Systems Command to form Air Force Materiel Command. Both of these actions signified that the constant pressures of Congress, the president, and DOD were forcing the Air Force to make some hard decisions on the best way to manage S&T for the future.

The establishment of four labs certainly amounted to a step in the right direction to show the Air Force's commitment to making progress in changing the organizational structure of its labs. But the Clinton administration, Congress, and DOD continued to pressure the Air Force to implement even more lab changes. In November 1993, President Clinton created the National Science and Technology Council, one of whose jobs entailed making a comprehensive study of military labs. As part of his plan to make big government smaller, Clinton sought to trim the size of the laboratory workforce and encouraged remaining employees to do more with less. The president also strongly favored a laboratory-restructuring plan that would lead to cross-service integration of resources—partnering with other DOD research agencies—to save money and boost organizational efficiency. Reaching this goal would require even deeper personnel cuts.

Passage of the National Defense Authorization Act in February 1996 prevented the Air Force from turning its back on making even more radical reforms to its laboratory infrastructure. This legislation directed DOD to develop a long-range strategic plan, known as *Vision 21,* that would spell out in very precise terms how the Air Force intended to consolidate into as few labs as practicable. The Air Force had to respond to the *Vision 21* tasking by coming up with a blueprint to revitalize its laboratory organizational structure—now beset by shrinking financial and personnel resources.

Two prominent individuals, Gen Henry Viccellio and Gen Richard Paul, were inextricably tied to the effort to reshape the laboratories. After assessing all the factors, they affirmed the inevitability, desirability, and timeliness of making fundamental changes to the lab infrastructure. General Paul, director of Science and Technology at Headquarters AFMC, emerged as one of the most influential players in the laboratory-reform movement, mainly because his organization had the most at stake in terms of alterations to lab infrastructure. Over the years, he had witnessed and could not ignore the disturbing and persistent pattern of decline in the number of people assigned to the labs. Pressures imposed by the Dorn cuts, defense planning guidance, and A-76 studies all contributed to significant personnel reductions. Paul realized that the current laboratory organization could not continue to absorb more and more personnel cuts without negatively affecting its mission. The handwriting was on the wall. Fewer people and fewer dollars convinced General Paul that maintaining one laboratory rather than four labs made more sense economically.

Paul's decision to propose a single lab was not a knee-jerk reaction. He was influenced over time by a number of strategic policy issues and high-level studies that consistently pushed for laboratory reform (fig. 10). George Abrahamson's blue-ribbon panel, *New World Vistas,* and the Air Force's *Global Engagement*

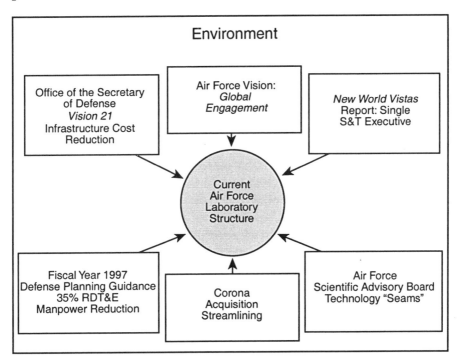

Figure 10. Issues Influencing the Creation of a Single Laboratory

vision all addressed lab issues that had an important impact on reorganization. The main thrust of these studies was that top Air Force leaders needed to look at rearranging the service's lab infrastructure to perform the missions of the twenty-first century. Everyone agreed that technology would remain a key ingredient in the success of future mission accomplishment, which meant fighting and winning wars. Consequently, the Air Force had to rebuild its lab infrastructure to develop and deliver the most advanced technology for supporting and protecting the war fighter.

With this in mind, General Paul proposed and General Viccellio approved the concept of one corporate laboratory as the optimum solution for changing the S&T infrastructure. Both men were convinced that one laboratory under the leadership of a single commander would result in more efficient control and integration of personnel, dollars, facilities, and technical programs. In addition, one lab would go a long way toward eliminating technology seams and reducing overhead by doing away with separate command sections and planning staffs.

By the summer of 1996, Paul and his staff were spearheading an intensive effort to sell the single-lab concept to the highest Air Force leaders. This involved preparing an acquisition issue paper for presentation at the Corona conference at the Air Force Academy in October. After several revisions, the Corona issue paper laid out the pros and cons of various options for reorganizing the labs. General Viccellio maintained that one corporate lab was by far the best option to meet the Air Force's future needs, and Secretary Widnall and General Fogleman agreed. Final approval to move forward with the implementation of the corporate lab occurred in the secretary's office on 20 November 1996.

The decision to create a single laboratory marked a pivotal turning point destined to have far-reaching consequences for the future of Air Force S&T. But future success for the Air Force depended to a large degree on its past performance in building a solid foundation for S&T. After World War II, Gen Hap Arnold and Dr. Theodore von Kármán had the foresight and wisdom to convince the military to take an active role in making S&T an integral part of the nation's long-range defense strategy. Over the past 50 years, Air Force laboratories have woven a rich legacy of lasting contributions to the nation's defense in such areas as nuclear and aerospace technology, development of intercontinental ballistic missiles, and revolutionary new weapon systems such as airborne lasers and high-power microwaves. In other areas, the labs have made tremendous progress with the modernization of bomber and tactical aircraft, the emergence of new and more durable materials, and the advancement of next-generation sensors and information systems. This extraordinary record continues to grow rapidly as S&T turns its attention to exploration beyond the atmosphere by developing more cost-effective and higher-performance space systems.

Looking back, one can easily tell that the Air Force laboratory system unquestionably has demonstrated its worth in sustaining the nation's defense. In the natural evolution of events, 13 labs in 1990 merged to four—six years later, they became one. This new Air Force Research Laboratory clearly represented the start of an exciting new era. It offered unlimited opportunity to make a lasting difference with the development of the world's most advanced technology that would provide the winning edge to the American war fighter. To make this happen, General Paul and his staff undertook the challenge of setting up the new consolidated lab as quickly as possible.

Part 2
The Transition

Chapter 9

Early Strategic Planning

General Paul's letter of 22 November 1996 to his four lab commanders, the director of the Air Force Office of Scientific Research, and his staff officially announced the secretary of the Air Force's approval of the proposal to create a single laboratory. This bold decision represented a major turning point in the Air Force's organizational structure and affected how the daily business of military science and technology would take place in the future. However, this grand scheme meant nothing unless plans and actions could be set in motion soon to transform four laboratories into one.

Bergamo

General Paul sensed the urgency and challenge of moving ahead promptly with the planning and implementation of the new lab. Above all, he knew it would take a total team commitment to succeed with the formidable undertaking that lay ahead: "We need to meet ASAP to begin the planning process for a single lab." As a first step toward getting this labor-intensive process under way, he directed his lab commanders, along with his key staff, to attend an off-site meeting scheduled for 5 and 6 December 1996 at the Bergamo conference center, located a few miles from Wright-Patterson AFB. Bergamo (formerly a monastery) offered the advantage of an informal atmosphere free from interruptions, which would allow everyone to furnish input to "corporately" build a phased transition plan leading to the stand-up of the single laboratory.[1]

From the vantage point of Air Force Materiel Command, General Viccellio also was very anxious to get the lab-planning process under way. From the start, his overall guidance to General Paul emphasized ensuring the achievement of certain single-lab "tenets" prior to the stand-up of the new organization. He reminded Paul that his first responsibility was to combine the four laboratories and AFOSR into a single lab under the direction of one commander and staff. In addition, Viccellio wanted Paul to reorganize the existing 22 technology directorates into roughly 10 to 12 large directorates. To integrate technologies across multiple directorates, he proposed establishing positions for three product directors—one each for Air Vehicles, C⁴I, and Space—who would report directly to the lab commander. In terms of personnel, Viccellio instructed Paul to begin planning to move all S&T employees to a new single-lab manpower document. Paul felt comfortable with the basic lab tenets laid out by Viccellio but wondered how he would develop an effective transition system to work out the "tons of details and decisions" that went along with setting up a single lab.[2]

Even though at this point, none of the technology directorates had been precisely defined, Viccellio, in an attempt to get the planning process off the ground, offered his vision of how the new organization should look (fig. 11). Viccellio's notional organizational chart mainly served as a compass to point everyone who was working on the lab's strategic planning process

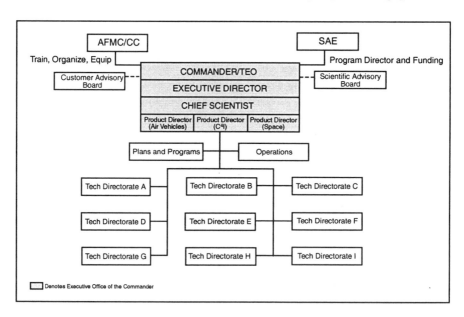

Figure 11. General Viccellio's Initial Vision of Single-Lab Organization, 26 November 1996 (From Gen Henry Viccellio Jr. to Distribution, letter, subject: Single Air Force Laboratory, 26 November 1996)

in the right direction. Not wanting to delay the formation of the single lab, he insisted that this priority project show progress quickly. Determined to see events move forward at a steady pace, Viccellio asked Paul to report back to him by mid-December.[3]

Knowing General Viccellio's expectations, Paul had less than two weeks to educate and organize his staff to begin addressing a number of complex issues that they would have to resolve before any new organization could stand up. Prior to the scheduled Bergamo meeting, Paul had met on several occasions with a small group of his closest advisors—Dr. Daniel, Mr. Dues, Dr. Russo, and Colonel Markisello—to brainstorm the best approach for molding a single lab from four diverse laboratories. What became painfully clear to Paul after only a few meetings was that it would take much more than the expertise and experience of this select group to assure that all the pieces would fall in place in a timely fashion. Paul knew almost immediately that he would have to depend on a much larger and heterogeneous group to make the lab happen. That was the first lesson learned in the entire lab-planning process:

> Actually, when this very small group got together, that's when I realized we needed a broader approach. We said, "Let's get three or four people, and we'll spend three or four days and map this out." What really became apparent to me was that was not the right way to do it. We needed much broader involvement, and we needed more time. We needed to bring more expertise in. That was invaluable in starting to lay out a [lab] structure. What hit me was we could not do this with a few people in a room. It is too complex—we wouldn't get the buy in. We just needed the diversity of more views and expertise on it. That's when we decided to bring a bigger group in.[4]

Paul's decision to bring a bigger group together was the main reason he arranged the Bergamo meeting for the first week of December. In the meantime, Paul and Dr. Russo had a very important meeting with Dr. John W. Lyons, director of the Army Research Laboratory, headquartered at Adelphi, Maryland, to find out firsthand how Lyons and his staff went about setting up their consolidated lab, formed in October 1992. They found out that the Army had taken 18 months to organizationally structure its single lab. But the critical lesson learned from Lyons was that the Army completed 90 percent of the solution for implementing its lab in approximately six months. With the benefit of hindsight, Lyons pointed out that he and his staff ended up spending the last 12 of the 18 months trying to solve the final 10 percent of their lab-consolidation problems.[5]

Lyons's perspective and comments made a strong impression on Paul, who left the meeting knowing he definitely did not want his transition team investing the majority of its time trying to solve the last 10 percent of lab-consolidation problems. He did not want to drag the process out because no matter the solution, it would not be perfect. Consequently, Paul's philosophy was to pursue the "90 percent solution," knowing full well ahead of time that they would make mistakes along the way—true of any new organization. General Paul believed it much more important to configure the lab quickly and make adjustments later, rather than waste time agonizing over making every piece of the lab puzzle fit exactly the first time around. In the end, this turned out to be a reasonable approach, since within 11 months of Secretary Widnall's approval, the four laboratories were inactivated and replaced by the Air Force Research Laboratory.[6]

Considering the geographic separation of the existing four laboratories and the magnitude of the operation required to establish a new lab, 11 months turned out to be a relatively short time to complete such an ambitious undertaking. Faced with a number of complex restructuring issues, the group that met at Bergamo ultimately had to take the lead in first identifying and then solving all the problems connected with the formation of a single lab. General Paul personally invited each person he had selected to attend the Bergamo meeting. He wanted his four commanders and AFOSR director present, as well as his experienced senior staff members, who he believed were strong corporate players. And since the secretary had already made the decision to go to a single lab, he was not interested in people debating the wisdom of that decision. It would be a waste of valuable time to fill the ranks of this all-important planning board with people not totally dedicated to the one-lab concept. Paul needed and sought out positive thinkers to discuss and develop a lab structure that would work. As it turned out, he had carefully assembled a highly motivated and qualified group determined to put together a smooth-working implementation plan. Those who attended this first critical planning meeting (table 5) represented a cross section of the existing organization and would play an extremely influential role in shaping the structure of the new laboratory.[7]

Table 5

Bergamo Meeting Team

Maj Gen Dick Paul	Dr. Don Daniel	Col Ron Hill
Dr. Helmut Hellwig	Col Rich Davis	Capt Deanna Won
Col Mike Heil	Col Ted Bowlds	Dr. Robert Selden
Mr. Terry Neighbor	Col Dennis Markisello	Dr. Joe Janni
Dr. Vince Russo	Dr. Earl Good	Mr. Ray Urtz
Capt Chuck Helwig	Ms. Wendy Campbell	Maj Mark Sabota
Mr. Tim Dues	Dr. Brendan Godfrey	

At the start of the Bergamo meeting, General Paul made a short but forceful presentation, explaining the impact of *Vision 21* on the evolution of the single lab. Although *Vision 21* called for fundamental changes in the organizational structure of the Air Force laboratory, he stressed that the mission of the lab would not change. The primary goal of the new organization was to continue to produce technology that would make the Air Force the world leader in the development of advanced weapon systems. Because that remained a constant, any organizational change had to conform to that goal. But he also pointed out that the mission would have to be accomplished with far fewer people, predicting that the number of workers assigned to the new lab would shrink to 5,507 by FY 2001 (down from over eighty-five hundred in FY 1989). He further acknowledged that the number of technology directorates would diminish and that a major focus in creating the new directorates entailed getting rid of technology seams by consolidating like technologies.[8]

In addition, developing and adhering to a strict schedule would be essential to complying with General Viccellio's overall guidance. Paul stated that they would have to put together a program plan to serve as a checklist of things to do before the lab could stand up. This would require a great deal of coordination and cooperation among the current lab commanders and their division, branch, and staff office chiefs, as well as Paul's immediate staff. For all this to succeed, Paul emphasized that he intended to implement a productive process for opening the lines of communications and fostering a free and healthy exchange of ideas and debate. As a start, he declared he would establish an AFMC Science and Technology Directorate web page to promote the sharing of information up and down the chain of command to "help quell the rumor mill." He felt very strongly about getting the word out and pledged to write the messages for the web site himself, rather than delegate this responsibility to his staff.[9]

During his closing remarks, General Paul offered a word of sober caution for the future. Although the single lab, he asserted, was the Air Force's answer to solving the very specific issue of intraservice lab reform, the new lab did not signal the end of the reform process. Rather, he realized that leaders at the highest levels of government were closely watching the outcome of the intraservice lab to determine if it made better sense to consolidate all service labs (of the Army, Navy, and Air Force) under an interservice plan some time in the future. That option always seemed to lurk in the background as a constant reminder of proponents at various levels of government who insisted that the combination of all service labs into one centralized DOD laboratory would be more beneficial economically over the long haul.[10]

The press also kept a watchful eye on the laboratory-reorganization effort. During the time of the Bergamo meeting, several highly charged articles appeared in the media, stressing that Congress was still considering legislation to strip the military labs of additional manpower and dollars. Excessive overhead at the labs remained a key issue that triggered civilian and government officials to reexamine the pros and cons of turning the labs over to the private sector or even closing them. Paul knew that, even though the Air Force had made its decision to form one lab, the press would continue to raise the issue of laboratory reform. To the Air Force leadership, the fact that the topic attracted such attention underscored the importance of making sure the transition plan was well thought out and executed in a timely fashion.[11]

Two days of meetings at the Bergamo center proved an important first step and a rewarding experience for the select group that brainstormed major laboratory-implementation issues. However, Paul made no final decisions. Instead, the meeting simply focused on starting an initial dialogue on a number of topics that would affect the stand-up and operation of the new lab. Everyone wondered how each of the technology directorates and the command section would be put together. After some debate, most agreed that AFOSR should remain a separate directorate rather than break up and distribute its basic research work among the various technology directorates. Beyond that, the exact organizational structure remained unclear.[12]

Obviously, identifying and structuring the tech directorates represented the most critical determinant of the success or failure of the lab. Someone had to decide how many tech directorates were needed, who would head them up, what technologies would be located where, whether some technology programs would have to move, and how many personnel would be in each directorate. These were complicated and sensitive issues, especially since the tech directors would undeniably become the highest-level leaders in the new organization. Consequently, General Paul asked Dr. Daniel to head a team to develop a concept for organizing the technology directorates. Its members included Mr. Urtz, Colonel Markisello, Mr. Dues, Captain Helwig, Dr. Russo, Mr. Neighbor, Dr. Good, Colonel Hill, and Ms. Campbell.[13]

Although formation of the tech directorates was a major concern, several other issues arose at the Bergamo meeting. For example, many believed that they should locate the new laboratory headquarters in Washington, D.C. Col Mike Heil, commander of Phillips Lab, and Dr. Brendan Godfrey, director of Armstrong Lab, led a spirited charge that strongly advocated the location of laboratory leaders in the nation's capital. There, they would be in a much better bargaining position to directly influence Congress and DOD officials on laws and policy affecting the laboratory. After all, the Naval Research Lab attributed much of its success over the years to its location in D.C., which allowed its leaders ready access to political and military decision makers. Paul felt that the new lab definitely needed a stronger presence in D.C., but he did not favor moving the headquarters there. Firstly, he did not want to isolate the headquarters from the largest component of the lab, located in Dayton, Ohio. Secondly, uprooting and moving east would entail tremendous physical and logistical problems, not to mention the enormous amount of lead time required to make a move of this magnitude. Thirdly, with D.C. real estate at a premium, where exactly would the headquarters locate? Setting up shop too far from the Pentagon and Capitol Hill (e.g., several miles outside the beltway) would defeat the whole purpose of colocation. As with the formation of the tech directorates, General Paul promised to commission a team to thoroughly assess the advantages of moving the headquarters to Washington, D.C.[14]

The Bergamo group also discussed possible names for the new lab. No one really wanted to commit to one name so early in the planning process. Nevertheless, it did not take long for the group to come up with a list: Air Force Research Laboratory, Air Force Air and Space Laboratory, Air Force von Kármán Laboratory, Aerospace Laboratory, Air and Space Laboratory, Air Force Research and Development Laboratory, Air Force Laboratory, and National Air and Space Laboratory.[15]

After the Bergamo meeting was over, General Paul continued to meet with his key staff regularly. Sometimes these were just one-on-one encounters so he could bounce ideas off his most trusted aides. Other times, he would meet with two to five people on his staff to get their input on specific aspects of the lab-planning process. These turned out to be useful meetings because they helped him prioritize strategic-planning issues. One of the first things he had to do was develop the briefing that Viccellio had requested by mid-December. Although it came two weeks late, on 31 December General Paul presented to General Viccellio the overall long-range plan for establishing the single lab.[16]

Phased Implementation

Paul's briefing to General Viccellio proposed a four-phased implementation approach that would start immediately and culminate nine months later with the stand-up of the Air Force single laboratory. Phase I, a three-month operation to be completed by the end of March 1997, called for the stand-up of the "interim" laboratory organization (fig. 12). Two major organizational changes would occur during this stage. Firstly, a single-lab command section and headquarters staff carved out of the existing 93-person AFMC/ST staff would increase to approximately 150 people who would serve as the nerve center to formulate and distribute overall lab policy and guidance. At that point, the command section would include the commander, executive director, chief scientist, and four product directors—for Air Vehicles, C⁴I, Space, and Human Systems. (Human Systems was added to General Viccellio's original suggestion of three product executives to provide parallelism with and identifiable linkages to the four existing product centers.) Most people assigned to the new single-lab staff would work either for XP or Operational Support (DS), where the support offices resided. The XP workers in each of the four

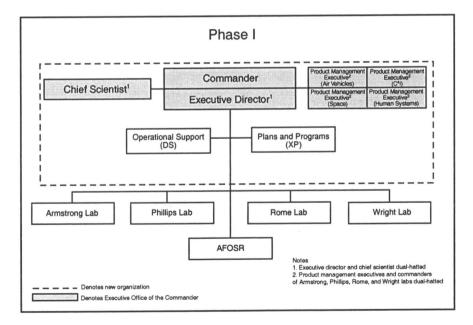

Figure 12. Phased Implementation Approach for a Single Laboratory (From briefing, Maj Gen Richard R. Paul, Headquarters AFMC/ST, subject: Single Laboratory Phased Implementation Approach, 31 December 1996)

labs and AFOSR would report to the single-lab XP to help formulate an integrated planning strategy. General Paul would direct the command section.[17]

A second major component of the phase I interim lab organization entailed a change in reporting procedures for the four lab commanders. No longer would they report to their respective product-center commanders but would report directly to the new AFRL commander, with the four labs' internal organizational structure remaining intact. This was a significant step forward because now the single-lab commander would have total control over personnel and the four commanders. One of the most notable benefits of this arrangement was that each lab commander no longer had to worry about serving two bosses—the TEO and the applicable product-center commander.[18]

Another important aspect of the role of the four commanders was that they would also become product directors in the single-lab command section. For example, Colonel Heil, the Phillips Lab commander, would become product director for Space; Col Rich Davis, who headed Wright Laboratory, would become product director for Air Vehicles; Col Ted Bowlds, head of Rome Lab, would become product director for C4I; and Dr. Godfrey, director of Armstrong Lab, would become product director for Human Systems. The product directors would help eliminate technology seams in the new lab by working to better integrate cross-directorate technologies for their assigned product or mission areas. Further, after the AFRL stood up, the lab-commander positions would no longer exist, and the product-director positions would become full-time. General Paul thought it important that the lab commanders occupy high positions of authority in the new lab structure to take advantage of their experience and expertise. As product directors, the four lab commanders not only would influence the success of the new organization, but also would be able to move into jobs that would not damage their careers after their old jobs went away.[19]

This new reporting system would allow General Paul to make maximum use of his four lab commanders during the critical implementation process. He thought it "essential that the four lab commanders work for [him] during phase I to help shape what the final phase should look like." Because people could have broken ranks during this critical time, it was imperative that the lab commanders work together to mitigate any of their employees' apprehensions about the new lab. Moreover, from a practical point of view, an element of uncertainty existed over the product commanders' somehow interfering with the lab-building process. General Paul also realized that the product-center commanders naturally did not relish the thought of losing control of either their laboratory commanders or their labs. Knowing he was about to lose his laboratory, a product-center commander conceivably could direct his lab commanders to move people out of the lab and assign them to the product center before formation of the single lab. To avoid that temptation, Paul campaigned aggressively to make sure that the four lab commanders came under his control as quickly as possible during phase I.[20]

Besides the four lab commanders, the director of AFOSR also would report directly to the single-lab commander as part of the phase I plan. Prior to approval of the single lab, AFOSR was classified as a field operating agency that reported to the Science and Technology Directorate at AFMC. However, as part of phase I, AFOSR would become a direct reporting unit—like the four labs. Moreover, the phase I organizational scheme called for the four labs and AFOSR to start a comprehensive review of every position assigned to their organizations—a necessary step in preparing to transfer all positions from the four labs and AFOSR to a newly created single-lab unit-manning document. Starting to build such a document was an extremely important exercise because it would consolidate lab personnel from the four product-center manning documents to one manning document, thus assuring that the single lab "owned" and controlled all government positions assigned to it. Other actions requiring resolution during phase I included choosing a name for the single lab, requesting an organization-change package (Air Force Instruction [AFI] 38-101, *Air Force Organization,* called for establishing the "interim" lab by the end of March), and appointing a transition team responsible for the entire implementation process.[21]

The value of phase I was that it would expeditiously establish a new interim laboratory with minimum disruption of the status quo. It would also pave the way for three subsequent phases. Paul explained to General Viccellio that the next step, phase IIA, would last about six months, ending in October 1997 (fig. 13). This would constitute a major milestone because by that time the single lab would have stood up with all current personnel positions from the four labs and AFOSR assigned to one unit-manning document. At that point, the four labs and AFOSR would cease to exist. Roughly 10 to 12 large technology director-ates would replace the four labs and AFOSR.[22]

Phase IIB, scheduled to take place from October 1997 to approximately FY 2001, would complete the "end-state" lab. This meant that the lab leadership would push hard to reduce the manning of the laboratory to conform to personnel numbers congressionally mandated by *Vision 21* and the DPG—a 35 percent reduction of lab personnel based on total authorized positions in 1989. Most of these reductions would occur in the

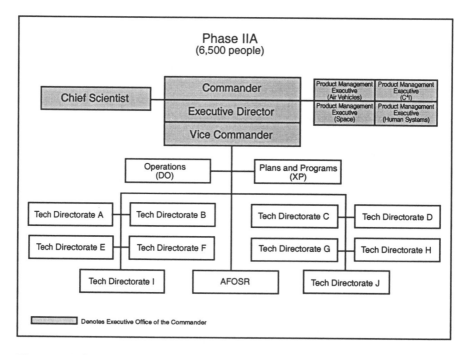

Figure 13. Second Stage of the Implementation Plan (From briefing, Maj Gen Richard R. Paul, Headquarters AFMC/ST, subject: Single Laboratory Phased Implementation Approach, 31 December 1996)

areas of support and overhead. Phase III, the final stage in the implementation process, looked to the lab environment nearly eight years into the future. The Air Force envisioned that by FY 2005, any lab closures and consolidations would be completed, and organizational adjustments would be made to comply with any interservice decisions resulting from *Vision 21*.[23]

Thus, phases I and IIA were the most immediate and important stages of the phased-implementation approach because those steps had to be completed before the lab stood up. After the lab was up and running, its leadership would be able to fine-tune and make more adjustments to the organization by reducing the number of personnel and closing or consolidating facilities, if necessary (and if permitted by subsequent BRACs), during phases IIB and III. This four-phased approach reflected General Paul's policy that called for implementation of the 90 percent solution by the end of phase IIA. The other "10 percent adjustment" could occur during phases IIB and III (fig. 14). Clearly, the most important priority was establishing the new lab and preparing it to operate on a day-to-day basis.[24]

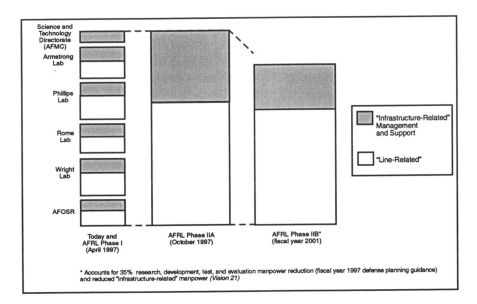

Figure 14. Evolution of the Single Lab (From briefing, Maj Gen Richard R. Paul, Headquarters AFMC/ST, subject: Single Laboratory Phased Implementation Approach, 31 December 1996)

After listening to all parts of General Paul's milestone briefing on 31 December, General Viccellio told him that he fully endorsed the phased approach for implementing the single Air Force laboratory. A week later, on 6 January 1997, Viccellio made it official by announcing that he, with the concurrence of the secretary of the Air Force for acquisition, had approved the phased approach and had directed Headquarters AFMC/ST to proceed with the implementation. Viccellio stressed that he wanted to move forward without delay and appealed to everyone in the organization to give complete support to this high-priority project. The next step called for General Paul to reconvene his staff in early January to begin hammering out the details of how to complete phases I and IIA.[25]

Transition Office

After General Viccellio gave the green light to move ahead with the implementation of the single lab, General Paul immediately formed a transition office to direct and monitor all activities related to the establishment of the new organization. During the weeks prior to Viccellio's approval of the phased approach, Dr. Daniel had already led a number of brainstorming sessions with Colonel Markisello, Vince Russo, Tim Dues, and others to begin to explore what procedures and processes would have to be put in place to get the new lab under way, especially the question of how to set up the technology directorates. These initial meetings took place in the conference room next to Dr. Russo's office at the Materials Directorate at Wright Lab. (After the lab stood up, Russo took a great deal of pride in referring to these early meetings in his conference room as the "birthplace" of AFRL.) However, not all the meetings occurred there. Earlier, over the Christmas holidays, Daniel had come in on a Saturday to meet with General Paul in his office, where they spent several hours discussing the importance of setting up a transition office as quickly as possible to resolve two key questions: how would the office be set up and who would run it?[26]

It became very apparent to Paul that Daniel could not realistically devote the time required to head the transition team.

Daniel already had a heavy workload and a very demanding schedule in his position as deputy director for S&T. One of his most important duties entailed dealing with internal, day-to-day S&T programs that still required managing, despite the existence of plans to radically change the laboratory's organizational structure. The business of the Science and Technology Directorate could not come to a halt. Consequently, Paul had to find a reliable person able to lead the transition team full-time. Fortunately, the general knew exactly the right person.[27]

On 13 January 1997, he selected Dr. Vince Russo to serve as the transition director. So that Russo could fulfill all the duties associated with his new position, Paul temporarily relieved him of his job as director of the Materials Directorate at Wright Laboratory, detailing him for 120 days to Headquarters AFMC/ST and assigning him an office just down the hall from the command section. This arrangement gave Paul ready access to Russo when he needed to consult with him on any aspect of the lab reorganization.[28]

General Paul was extremely impressed with Russo's experience and proven abilities. From the beginning, Paul believed it important to have a person in the Senior Executive Service to lead the transition office, a requirement that Russo met. Moreover, Paul had known Russo personally and had worked with him over the years on a number of lab projects: "I highly respected his opinion very, very much. He is a strong leader and team builder, and I also felt he had credibility with the other lab commanders. And he knew how to build teams. I watched him do it over and over within his own organization." After two months on the job, Russo earned Paul's praise as a man who "ha[d] 'lab blood' running through his veins" and who completely understood the workings of the Air Force laboratories.[29]

Paul particularly liked Russo's management style. He neither constantly criticized nor worked in isolation. Rather, he actively sought the opinions of others, urged people to get involved, and made them feel that their ideas were important and made a difference—exactly the sort of leader Paul wanted. He was certain that Dr. Russo would tackle the transition problem head-on by using a large cross section of the lab workforce to come up with workable solutions. Up until this point, only a handful of people—by design—had participated in the transition planning, but Paul wanted to change that because too many questions remained unanswered under that system. In his mind, no consensus existed on even the most basic issues. Russo offered a refreshing new approach that relied on a more "participatory" methodology to ensure the success of the lab-transition plan. Because of Russo's personality, experience, interpersonal skills, and leadership abilities, Paul had supreme confidence in his ability to serve as the transition director.[30]

Paul was equally certain that Russo could not single-handedly make the new laboratory happen. It was far too big a project. So he made sure Russo had adequate full-time help by assigning about half a dozen people to the transition office. To underscore the importance of the enormous task ahead, General Paul approved the lab-transition structure on the same day he appointed Russo director (fig. 15).[31]

General Paul firmly believed that the transition office would serve as the backbone of a successful implementation program, and he depended on the transition team to develop a single-lab planning process that eschewed quick solutions. He insisted on an open process that carefully considered "all the

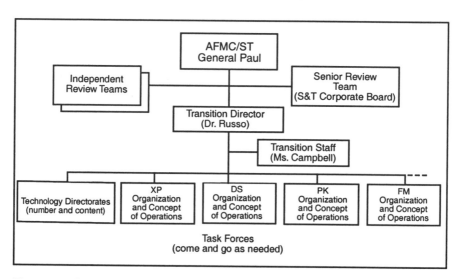

Figure 15. Single-Lab Transition Structure (From Maj Gen Richard R. Paul to AL/CC et al., letter, subject: Transition Director for Single Laboratory, 13 January 1997, with attached chart "Single Lab Transition Organization")

alternatives and document[ed] our rationale for why we rejected some options and pursued others." He wanted the team to encourage a wide spectrum of the current lab workforce to participate in developing the lab-implementation plan: "Without such a transition organization populated by a cross-section of our workforce, we would tend to have a 'closed process.' That's the antithesis of what we want." The appointment of Ms. Wendy Campbell as deputy transition director exempli-fied Paul's policy of recruiting people from different back-grounds to serve on the transition team.[32]

Ms. Wendy Campbell served as a key member of the lab-transition team.

A research psychologist rather than a scientist, Wendy Campbell had worked in the Human Resources office at Armstrong Laboratory in San Antonio, where she developed her skills as a highly proficient manager in working "people" issues. General Paul had selected her to come to Wright-Patterson to plan and implement the Lab Demo program in 1996. An innovative civilian-appraisal system, Lab Demo evaluated each scientist and engineer's (S&E) contribution, commensurate with his or her grade level, and then tied salary increases to the S&E's contribution score. Outsiders immediately embraced this radical approach for appraising civil servants as visionary and bold. Campbell had done an exceptional job with Lab Demo, exceeding everyone's expectations. In the process, she had earned a reputation as a self-starter with a tremendous amount of energy, drive, and stamina to get things done on time. As a result, General Paul asked her to become Vince Russo's deputy and attend to all the day-to-day details of running the lab-transition office.[33]

Paul often referred to Russo and Campbell as the "one-two punch" of the transition office. Although probably the two most visible individuals in that office, they depended heavily on the other staff members, who made invaluable contributions. These tenacious workers in the trenches interacted daily with the various task groups charged with organizing the major components of the new lab. Consistent with General Paul's policy of drawing people from throughout the organization, these staffers came from Headquarters AFMC/ST and each of the four laboratories for a four-month assignment to the transition team (table 6).[34]

Table 6

Air Force Single-Lab Transition Staff (January 1997)

Name	Organization	Task Group Affiliation
Dr. Vince Russo	Materials Directorate (ST/SL)	All
Ms. Wendy Campbell	Materials Directorate (ST/SL)	All
Dr. Harro Ackermann	Phillips Laboratory	Product Executives and Financial Management
Maj Jack Donnelly	Plans and Programs (ST/XP)	Contracting and Corporate Information
Mr. Don Elefante	Rome Laboratory	Personnel and Washington Presence
Capt Charles Helwig	Plans and Programs (ST/XP)	Technology Directorates and Plans and Programs
Mr. Thomas Hummel	Wright Laboratory	Headquarters Locations and Operational Support
Capt Scott Jones	Armstrong Laboratory	Integration and Operations Division

Source: Lab transition team, chart, "Air Force Single Lab Transition Staff," 11 February 1997; and idem, chart, "Single Lab Transition Team Points of Contact," 26 March 1997.

After naming the transition staff, General Paul and Dr. Russo then selected 13 task-group leaders and the areas they were to address. Paul had some very definite ideas about the composition of the task groups' working membership. Here too, he wanted a mixture of senior people, each of whom had a great deal of experience and expertise in one of several broad areas of lab operations. But he also insisted that some members represent a cross section of the workforce at large, thus giving less experienced people an opportunity to inject new ideas into the strategic-planning process. Although there were no hard-and-fast selection rules, Dr. Russo met with each task-group leader to identify group membership. To make this work, he gave each leader the authority to write each of the four lab commanders and the AFOSR director requesting that they provide a capable and interested person—who could think corporately—to serve on the various task groups. (The commanders and director were not allowed to prevent people from serving.) Once selected, that person would work directly with the task-group leader and other members of the group, who would meet periodically at the discretion of the group leader (table 7).[35]

The most important responsibilities of the transition office's staff included providing overall leadership and oversight by identifying, scheduling, coordinating, and monitoring the activities of a number of task groups and focus teams, the latter addressing specialized concerns outside the purview of the task groups. Each focus group would consist of "experts from the field" who would resolve very specific issues affecting the stand-up of the new laboratory. Phase I included six focus

Table 7

Task Groups and Leaders

Task Group Leader	Task Group	Organization
Mr. Tim Dues (SES)	Plans and Programs (XP)	Wright Lab/XP
Col Dennis Markisello	Support (DS)	AFMC/ST
Mr. Rich Eckhardt (SES)	Financial Management (FM)	AFMC/FM
Maj Gen Rich Roellig	Contracting (PK)	AFMC/PK
Mr. Lief Peterson (GM-15)	Personnel	AFMC/DP
Brig Gen (sel) Rich Davis	Product Executive Officers	Wright Lab/CC
Mr. Garry Barringer (SES)	Corporate Information	Rome Lab/XP
Dr. Brendan Godfrey (SES)	Washington Presence	Armstrong Lab/CC
Col Walt Avila	Classified Programs	AFMC/DR
Col John Rogacki	Integration and Operation Divisions	Phillips Lab/RK
Dr. Vince Russo (SES)	Headquarters Location	AFMC/ST-SL
Dr. Bart Barthelemy (SES)	Technology Directorates	Retired
Brig Gen Charles King	Reserves	Reservist AFMC/ST

teams dealing with the following areas: enlisted-personnel issues, laboratory-heritage preservation, administrative processes, chief-scientist functions, command issues, and AFRL regulations/policies/instructions/operating instructions.[36]

Each task-group leader would act as the facilitator and spokesman for his group, each of which sought to furnish information on how the new lab should organize and function. Group members were to brainstorm a wide range of options for their assigned "broad" area in an attempt to come up with a few logical choices and document the rationale for their selections. After making an objective list of pros and cons for each option, the group would recommend the best option to help complete the overall implementation process. Thus, each group would come up with one of many pieces for integration into the total lab puzzle.[37]

Each task group would present its findings in an issue paper to the Science and Technology Corporate Board, consisting of General Paul; his executive director, Dr. Daniel; the four lab commanders; the AFOSR director; and selected members of Paul's senior staff. Traditionally, this board had addressed long-term technology processes/products and investment strategies that required corporate-level decisions. After the board's review, two other specially constituted review boards would then evaluate each task group's recommendations and offer their opinions on the soundness of each option. The external Independent Assessment Board (IAB) consisted of senior people from industry, academia, and government, while members of the internal Grassroots Review Board represented a cross section of the current laboratory system's S&T workforce. As part of this process, each board would identify flaws in the logic of each task group's recommendation, raise other issues that might have been overlooked, and call the Corporate Board's attention to potential political implications. After listening to and evaluating all the facts presented by the task group and the two boards, Paul would solicit final advice from the Corporate Board and then either approve or disapprove each recommendation. Once Paul accepted a recommendation as the best solution, it then became part of the transition plan for people to follow during phase IIA.[38]

All of the task groups—which General Paul referred to as "the heart of the transition structure"—faced a formidable challenge, considering the very short time they had. Indeed, not until mid-February were all members of the task groups selected. Group leaders then acted quickly to set up a meeting in which all members could participate—a seemingly simple task that often took a great deal of coordination because members were literally spread out from coast to coast. As a result, most of the groups' kickoff meetings did not occur until the end of February or early March, leaving little time to formulate recommendations to brief to the Corporate Board by the third week of March. By the end of March/early April, the schedule called for the IAB and Grassroots Review Board to start collecting input to pass on to lab leaders for better refining their strategic game plan.[39]

Stand-up of the interim laboratory at the end of March would complete phase I but did not mean that General Paul and the Corporate Board could finalize the recommendations of the task groups. That ongoing process would continue until early July. Paul and others expected that the initial inputs from the task groups would most likely require changes and fine-tuning in order to establish agreed-upon procedures for the final push in the summer of 1997, prior to October's stand-up. So the strategy called for each task group to rework its initial input and develop a revised plan that considered suggestions from the first meeting of the Corporate Board. Each group would present its final product to one of several Corporate Board meetings held between the end of April and early June. During the first two weeks of July, Paul intended to make a final decision on the groups' recommendations. Although the general acknowledged the aggressiveness of this timetable, especially considering the scope and complexity of the work involved, he insisted on following his "90 percent rule" to set up the lab as quickly as possible. Paul repeatedly reminded group leaders and members of their jobs' "highest priority" and encouraged them to complete their work on schedule. They did not disappoint the general.[40]

Notes

1. Maj Gen Richard R. Paul, director, Science and Technology, to Distribution, letter, subject: Single Lab Planning Offsite, 22 November 1996.

157

2. Gen Henry Viccellio Jr. to Distribution, letter, subject: Single Air Force Laboratory, 26 November 1996; and Maj Gen Richard R. Paul, interviewed by author, 2 March 1998.

3. Viccellio letter.

4. Maj Gen Richard R. Paul, interviewed by author, 22 November 1999.

5. Ibid.; and Capt Chuck Helwig, AFMC/ST (advanced technology integration manager), interviewed by author, 3 February 1998.

6. Paul interview, 22 November 1999; and Helwig interview.

7. Wendy B. Campbell, Single Laboratory Secretariat, "Summary of the Single Laboratory Confidante Review," report [December 1996]; and Paul letter.

8. Col Dennis Markisello, "Single Lab Offsite," notes prepared at the Bergamo meeting, 5–6 December 1996.

9. The web site was not operational until 11 March 1997. Since anyone could access the original site, Public Affairs had to approve it—a process that took time and created delays. As an interim measure, the web site was restricted to ".mil" users so General Paul could get the word out quickly. Ibid.; and Paul interview, 2 March 1998.

10. Markisello; and Paul interview, 2 March 1998.

11. Jeff Erlich, "Pentagon Drafts Lab-Cutting Legislation," *Defense News,* 2–8 December 1996, 1, 34; and John Pulley, "Labs May Be Next Closure Prospect," *Federal Times,* 9 December 1996, 6.

12. Maj Gen Richard R. Paul to all Headquarters AFMC/ST, Laboratory, and AFOSR employees, letter, subject: Update on Single Air Force Laboratory, 17 December 1996; and idem, "Single Lab Offsite," notes prepared after the Bergamo meeting, 5–6 December 1996.

13. Markisello; and Ms. Wendy Campbell, interviewed by author, 11 June 1998.

14. Campbell, "Summary of the Single Laboratory Confidante Review."

15. Ibid.

16. Briefing, Maj Gen Richard R. Paul, Headquarters AFMC/ST, subject: Single Laboratory Phased Implementation Approach, 31 December 1996.

17. Ibid.; Dr. Vince Russo, interviewed by author, 4 February 1998; and Dr. Vince Russo, "Functional Support in AFRL," report, 5 June 1997.

18. Paul briefing.

19. Gen Henry Viccellio Jr. to Distribution, letter, subject: Single Air Force Laboratory, 6 January 1997; and Paul interview, 22 November 1999.

20. Paul interview, 22 November 1999; and Helwig interview.

21. In terms of the personnel system, the organizational-change request permitted the four labs to report to another laboratory—a temporary arrangement because in October 1997 the four labs would go away and only the single lab (AFRL) would remain. Paul briefing.

22. Ibid.

23. Ibid.; Viccellio letter, 6 January 1997; and briefing, Maj Gen Richard R. Paul, subject: Air Force Research Laboratory: Heading Check, 29 April 1997.

24. Paul briefing, 31 December 1996; Viccellio letter, 6 January 1997; Paul briefing, 29 April 1997; and Paul interview, 22 November 1999.

25. Viccellio letter, 6 January 1997.

26. Dr. Don Daniel, interviewed by author, 27 July 1998; Paul interview, 2 March 1998; and Russo interview.

27. Daniel interview; Paul interview, 2 March 1998; and Russo interview.

28. Maj Gen Richard R. Paul to AL/CC et al., letter, subject: Transition Director for Single Laboratory, 13 January 1997.

29. Paul interview, 2 March 1998; and message, General Paul's web site, subject: Single Laboratory Transition Organization; on-line, Internet, 18 March 1997, available from http://stbbs.wpafb.af.mil/STBBS/labs/single-lab/updates.htm.

30. Paul interview, 2 March 1998.

31. Paul letter, 13 January 1997.

32. Paul message.

33. Ibid.; Paul interview, 2 March 1998; and Campbell interview.

34. Lab transition team, chart, "Air Force Single Lab Transition Staff," 11 February 1997; idem, chart, "Single Lab Transition Team Points of Contact," 26 March 1997; Paul interview, 2 March 1998; and Dr. Harro Ackermann, AFRL/DEL, interviewed by author, 10 January 2000.

35. Ackermann interview; and Maj Gen Richard R. Paul, to AFMC/FM et al., letter, subject: Task Group Leaders for the Single Laboratory Transition Activities, 14 February 1997.

36. Paul letter, 14 February 1997; briefing, Dr. Vince Russo, subject: Kickoff Meeting: Single Laboratory Contracting Task Group, 19 February 1997; message, General Paul's web site, subject: Single Lab Task Groups and Focus Groups; on-line, Internet, 23 March 1997, available from http://stbbs.wpafb.af.mil/STBBS/labs/single-lab/updates.htm; and Campbell interview.

37. Paul letter, 14 February 1997; and Russo briefing.

38. Paul letter, 14 February 1997; message, General Paul's web site, subject: 20–22 March Corporate Board Meeting; on-line, Internet, 20 March 1997, available from http://stbbs.wpafb.af.mil/STBBS/labs/single-lab/updates.htm; and briefing, Maj Gen Richard R. Paul, subject: Air Force Single Laboratory, 13 February 1997.

39. Russo briefing; and Paul message, 18 March 1997.

40. Russo briefing; Paul message, 18 March 1997; Paul interview, 2 March 1998; and message, General Paul's web site, subject: AFRL Phase II: A Decision Process Flowchart; on-line, Internet, 25 April 1997, available from http://stbbs.wpafb.af.mil/STBBS/labs/single-lab/updates.htm.

Chapter 10

Shaping the Technology Directorates

One issue, more than any other, drove the implementation plan leading to the stand-up of the lab: reorganizing and consolidating the large number of existing technology directorates into a smaller number of new directorates. Science and technology work, the center of the laboratory operation, took place in the technology directorates. The tech directorates affected almost every aspect of the laboratory, including contracting, support staff, command guidance, personnel numbers, funding, and more. Thus, General Paul was acutely aware of the importance of selecting a highly competent tech-group leader who could work effectively with the existing tech directors and other senior leaders. He would have to be a savvy person with sufficient street smarts to guide his task group down a rocky path and then determine how many new tech directorates would be needed, what technologies they should represent, and where they would be physically located within the single-lab structure. In other words, the basic question became, How do you transform 22 tech directorates into some smaller number that would exhibit both "technology purity" and critical mass? Paul did not know the exact answer to that question. That is why he was anxious to hire someone soon to solve what he described as the "hardest piece by far" of the entire lab-reorganization puzzle.[1]

As it turned out, Vince Russo played a very influential role in the selection of Dr. Robert R. Barthelemy to lead the Technology Directorate Task Group. Earlier, Russo had met with General Paul to discuss the qualifications a person would need to head such an important undertaking. After lengthy discussion, Paul and Russo agreed that whomever they chose had to be a good facilitator above everything else: "We were looking for a guy who would be dealing with controversial subjects. Number one, we needed a facilitator—someone who is easy, someone who can get along with lots of people . . . someone who is able to get others to get along and work together toward a common goal." The ideal role of the facilitator was not to make the final decisions but to persuade others, through a give-and-take process, to reach a consensus on the best way to organize the tech directorates. Russo had known Barthelemy for over 25 years and had worked with him on various lab projects. Over that time, Russo concluded that Barthelemy was one of the most competent laboratory professionals he had ever encountered in his 35-year government career.[2]

General Paul concurred enthusiastically with Russo's assessment of Dr. Barthelemy. Although Paul did not know Barthelemy as well as Russo did, Paul had worked with Barthelemy on several occasions during his days at Wright Laboratory and was impressed by his aptitude for getting people to work together and accomplishing tasks quickly. He needed a person who could charge ahead in an orderly fashion and meet deadlines. In addition, the tech-group leader would have to deal with a diversified group of current tech directors who held very strong and different opinions on the way the lab should reorganize. After Barthelemy retired in January 1996, he went to work for Universal Technology Corporation in Dayton, Ohio, where he became involved in consulting a number of private companies. In that role, he developed management and leadership skills by conducting numerous workshops and seminars and earned a reputation as an excellent facilitator who worked with company officials and employees to make improvements in their daily operations.[3]

Furthermore, Barthelemy knew how the laboratory system functioned at Wright-Patterson, having worked as an Air Force officer and civilian at the various labs there since 1963. He had held high-level positions over the years and had gained a great deal of experience working a variety of S&T programs. Most importantly, he had worked his way up through the laboratory system over the years. Early on, he served as an engineer in the Aero Propulsion Lab's Plans Office. In the 1970s, he had assignments in the areas of high-power lasers and the basic mechanics involved in heat pipes. In 1982 he became the chief of XP at the newly formed Air Force Wright Aeronautical Laboratories. The next year, he became a member of the SES and assumed the position of deputy director of the Aero Propulsion Laboratory. Before he retired from federal service in January 1996, he spent seven years running the

National Aerospace Plane program and three years as director of the Training System Program Office at AFMC's Aeronautical Systems Center in Dayton.[4]

Knowing that Barthelemy had a strong lab background made a big difference to General Paul. Having worked in the labs made Barthelemy an ideal candidate because he could speak "lab talk," thereby catching subtle nuances that could easily escape a less experienced person. Moreover, the current tech directors could not easily dismiss him, given his credibility and understanding of how the labs operated. Moreover, the fact that Barthelemy was now

General Paul selected Dr. Robert "Bart" Barthelemy to lead the all-important Technology Directorate Task Group.

a "lab outsider" since he had retired and brought no agenda to the table made him even more appealing to Paul because he could now objectively assess how the labs should be restructured in terms of technology disciplines. Paul had no apprehensions at all about putting Barthelemy in charge of the all-important Technology Directorate Task Group.[5]

Barthelemy confirmed that Russo had approached him in late December and again in early January to tell him that General Paul was looking for someone on the outside who did not have a stake in the outcome of the reorganization of the directorates. Russo and Paul anticipated some knotty problems because the current directors would most likely have a narrow perspective and an unyielding attitude about protecting their technology turf. Consequently, Paul preferred to bring in someone who could open everyone's eyes to the total corporate picture and convince the group members to compromise on key issues as they sought to effectively organize the new laboratory.[6]

Barthelemy was in California working on a consulting job when he received a phone call from General Paul in the middle of January asking him to help out with the restructuring of the technology directorates. He told Paul he was interested in the position and flew back to meet with him during the last week of January to discuss his duties and responsibilities. Paul told Barthelemy he wanted him to focus on reducing the 22 technology directorates at the Phillips, Wright, Rome, and Armstrong labs to eight to 12 new directorates. Barthelemy and his group were to study the issue and come up with a recommendation for the number and kinds of tech directorates.[7]

Secondly, Paul directed Barthelemy to regroup the technology directorates in a way that would minimize technology seams that permeated the existing laboratory organization. In electronics, for example, bits and pieces of work took place in seven or eight different directorates—a practice that many people, including Paul, considered redundant and a duplication of effort. The general asked Barthelemy to look for realistic ways of consolidating the majority of electronics work under one or two tech directorates.[8]

Reducing the number of technology directorates and consolidating similar work efforts into a single directorate would also reduce the number of subordinate units, which included division and branch levels. Fewer tech directorates automatically translated to fewer division and branch chiefs. As a third goal for Barthelemy, Paul favored reducing the number of these middle-management positions in accordance with the overall plan of setting up the new lab as a flatter organization with fewer management and support positions. However, this did not mean that middle managers would lose their jobs. Instead, the plan called for them to move back into the various tech divisions and branches as program managers or bench scientists. Paul considered this a strength of the new laboratory because more scientists and engineers would become directly engaged in working the core S&T issues destined to advance systems that would best support the war fighter.[9]

Paul allowed Barthelemy to pick whomever he wanted to serve on the group, specifying only that the members should represent all four labs and AFOSR. To launch the process, Paul assigned Col Ron Hill from the Human Resources Directorate at Wright Lab to serve as Barthelemy's deputy, as well as Capt Chuck Helwig from the general's command section, Capt Jeff Witco from AFMC's Requirements Directorate, and

Ms. Elona Beans from Armstrong Lab at Wright-Patterson to provide administrative support. By early February, Barthelemy, who wanted to keep his group relatively small, had identified 12 people to serve on the task group (table 8).[10]

Table 8

Technology Directorate Task-Group Membership

Wright Lab	Rome Lab	Armstrong Lab	Phillips Lab
Dr. Alan Garscadden	Dr. Donald Bodnar	Dr. Russell Burton	Ms. Christine Anderson
Dr. G. Keith Richey	Mr. Igor Plonisch	Dr. David Erwin	Col William Heckathorn
Mr. Steve Korn	Dr. John Granier	Dr. Lee Task	Mr. Joe Sciabica

Barthelemy went to work right away, calling the first meeting for 25–26 February at the Hope Hotel across the street from Headquarters AFMC at Wright-Patterson. In preparation for this kickoff meeting, he pointed out to all the team members that "there is no approved solution coming into this meeting." Instead, he wanted people to come with open minds and a willingness to brainstorm a number of options for shaping the new tech directorates in ways most beneficial to the entire organization. Although Barthelemy explained that he would not follow a hard agenda because he wanted the group to collectively make its decisions, he did identify a number of concepts to be addressed at the meeting: discussion of goals, criteria, and ground rules; General Paul's guidance; posturing AFRL for the future; and process definition and development.[11]

Barthelemy realized that his group, which he referred to as "The Twelve," represented only half of the four labs' 22 tech directorates. However, the 12 members did represent all four labs, as General Paul had requested, and seemed a more manageable number than 22 for getting things done and for engaging in "team building" activities within the specified time constraints. His group had less than one month to brainstorm, develop, and analyze various options; decide on the two best options for each directorate; and prepare its findings for presentation to General Paul and the Corporate Board meeting scheduled for 20–22 March.[12]

During General Paul's opening remarks to the task group members on 25 February, he made it very clear that he depended on them to develop the best possible plan for organizing the tech directorates under one laboratory. He told them he had no doubt that they would lay the foundation for the new laboratory and that almost every other lab-reorganization activity would hinge on their plan for setting up the new tech directorates.[13]

Explaining his preference for eight to 12 new directorates, Paul noted that a greater number would continue the problem of having technologies "fragmented" over too many directorates and that a lesser number would create very large directorates that would be difficult to manage and control. However, he pointed out that if the group came up with a different number that made sense, he would support it. Paul was also enthusiastic about the caliber of people assigned to the group, reminding them that they had a rare opportunity to develop a meaningful plan that would ultimately change the future of the laboratory system in the Air Force. Before he departed, Paul expressed his complete confidence in their ability to provide an extremely valuable and lasting service to the new lab and to the Air Force. He also assured them that he would make himself available and provide whatever support he could to ensure the success of their mission.[14]

Although extremely pleased with Paul's comments and the urgency he attached to the group's challenging job, Barthelemy realized that getting all 12 members to think corporately would be no easy job. He characterized them as extremely bright and capable individuals who were very competitive, aggressive, independent, and strong-willed with big egos. His principal chore involved harnessing and channeling all their talents and energy in the same direction to allow them to determine the optimum grouping of technologies in the new lab's directorates. To facilitate their understanding of the job they faced, Barthelemy borrowed a management technique he had used in consulting work for some of the largest private companies in the defense industry:

> All 12 of them came in ready to fight for what they wanted. It turns out that there were about 24 [22] directorates and about a hundred divisions [115] within those tech directorates. So I went to the store and

bought jelly beans. I got 24 different kinds, different colors . . . four or five of each color representing the hundred divisions and the 24 tech directorates. On the first day, I threw the jelly beans on the main table, and I said, "This is our problem. We've got to arrange these jelly beans." And I said we can spend the next six months putting together the yellows and reds and all that. Or we can just eat these jelly beans and start talking about what's right for the organization.[15]

This little icebreaker not only reduced tensions but also emphasized that the group should not think in parochial terms. For the good of the lab, the members had to expand their perspectives. As the group became more immersed in the problem-solving process, everyone realized that no one would get exactly what he or she wanted and that their first priority was the welfare of the new laboratory. In time, all of the group members saw the wisdom in that corporate approach and rallied to put into practice the Air Force core value of "service before self."[16]

The most important point considered at this first meeting was the aforementioned goal of reducing the number of existing tech directorates and providing the Corporate Board several organizational options that would meet the objectives and constraints of the new Air Force Research Laboratory. In developing options, Barthelemy instructed his group to take into account a number of pertinent objectives: minimizing technology seams, enhancing core competencies, streamlining the directorates, maintaining best practices, enhancing customer orientation, establishing directorates of three hundred to seven hundred personnel and divisions of one hundred to three hundred personnel, and considering AFOSR as a directorate. In terms of constraints, the group could not move any of the directorate components and could use only current laboratory resources—nothing from outside organizations (e.g., people, funding, facilities, etc.). In other words, the four laboratories + AFOSR + ST = AFRL.[17]

Once the 12 task-group members understood these ground rules, selection of the tech directorates progressed smoothly and more rapidly than expected. During the first round of discussions, the group quickly saw that certain existing technology directorates easily lent themselves to integration into a new and larger directorate under the single lab. In addition, the members realized that many of those existing directorates

(Barthelemy estimated 80 percent) could logically merge with a minimum of debate. Accordingly, the members agreed that they could group existing technologies under separate directorates for work conducted in materials, space, directed energy, air vehicles, information, and human effectiveness (including health, safety and environment, and crew effectiveness).[18]

Although these new directorates essentially accounted for the same technologies as in the past, the difference was that bits and pieces of similar technologies—previously scattered among several directorates—would now be consolidated into one larger tech directorate. For example, under the four-lab system, the Geophysics Directorate studied and defined the effects of the upper atmosphere and space environment on aerospace systems. Since geophysics work focused on space, the task group reasoned that it made sense to eliminate the existing Geophysics Directorate and move its work into the new space directorate. Also, General Paul thought that, despite its importance as a scientific discipline, geophysics would not grow substantially in terms of attracting additional resources in the future. For these reasons, the group found it difficult to justify geophysics as a separate directorate and chose to merge it with the Space Vehicles Directorate.[19]

Other cases were just as easily resolved. For example, no one questioned that AFOSR, which managed basic research programs across the laboratory system, would become a separate directorate. Essentially, this office would retain its identity and continue to operate as it had in the past. Barthelemy's group also proposed formation of a single munitions directorate responsible for air-launched weapons designed to defeat ground/fixed, mobile/relocatable air and space targets. Although this seemed the right approach for keeping all work on conventional weapons together, the group also considered integrating certain parts of munitions work for inclusion in the Directed Energy Directorate. However, after concluding that the technologies of directed energy and conventional weapons were distinctly different, the group favored keeping munitions work intact under a separate directorate. By quickly identifying most of the new tech directorates that would become an integral part of the new laboratory, the group gave itself more time to consider the more difficult organizational restructuring that focused on establishing roughly 20 percent of the remaining new tech directorates.[20]

Less certainty and more controversy attended the remaining technology directorates. Everyone knew that Air Vehicles and Space Vehicles would be important directorates in the new lab but wondered whether each should have a propulsion component. Col John Rogacki, who headed Phillips Lab's Propulsion Directorate at Edwards AFB, California, and Dr. Tom Curran, his counterpart at the Aero Propulsion and Power Directorate at Wright-Patterson, strongly advocated the establishment of a separate Propulsion Directorate under the new lab. Because of the natural synergy among air, rocket, and space propulsion, as well as the prominence of propulsion as an area of technology, they argued against moving air-breathing propulsion to Air Vehicles and rocket propulsion to Space Vehicles. To them, assigning propulsion to two different directorates would only create more technology seams in the organization, which violated General Paul's policy of eliminating such seams.[21]

In addition, over the past few years, Rogacki had concluded that his Propulsion Directorate had become a stepchild of Phillips Lab, mainly because of the geographical separation between Kirtland AFB and Edwards AFB. The impression, whether right or wrong, was that the Phillips hierarchy tended to take better care of its local directorates at Kirtland (especially when it came to personnel and funding cuts) than its Propulsion Directorate, located several hundred miles away from the Phillips Lab headquarters in Albuquerque. On many occasions, Rogacki and his rocket-propulsion proponents felt they received less than a fair shake from Kirtland. As a result, the propulsion people decided they would rather take their chances operating as a separate directorate at Wright-Patterson, where they would be in a better position to take control of their destiny.[22]

On the other hand, Christine Anderson, who led the Space Technology Directorate at Phillips Lab, and Dr. Keith Richey, director of the Flight Dynamics Directorate at Wright Lab, felt quite the opposite. They led a subgroup to collect input from people in their respective divisions and branches on the advantages and disadvantages of establishing a separate Propulsion Directorate. Not surprisingly, they disputed the notion of limiting all propulsion work to one directorate. Anderson stated that because the launch-propulsion element of lifting missiles off the ground would become increasingly dominated by the commercial sector, it was not her primary concern for the future of space technology. She had more interest in developing advanced technology dealing with onboard propulsion systems for spacecraft. Although a less mature technology than propulsion for launch vehicles, onboard propulsion had far more potential in the near future for influencing higher performance levels in terms of the on-orbit precision movement of spacecraft.[23]

Barthelemy's group found itself unable to choose between establishing a stand-alone Propulsion Directorate or combining parts of propulsion into the Space Vehicles and Air Vehicles directorates. Neither was it able to choose after hearing the subsequent findings of Anderson and Richey's subgroup. At that point, the group simply decided to provide a set of possible alternatives to General Paul and the Corporate Board in March. Since the group had reached a stalemate, Barthelemy, as group leader, decided to break the tie and recommend a separate Propulsion Directorate at the Corporate Board meeting in April. Barthelemy's view prevailed—by June, General Paul had elected to establish a separate Propulsion Directorate rather than split that work between space and air vehicles. Despite their disappointment, Anderson and Richey, as team players, accepted the decision. Anderson remarked, "We will make it work."[24]

Later in the summer, concerning a related restructuring issue, Anderson objected to the name given the proposed Space Directorate she would lead. In her mind, *Space Vehicles Directorate* was far too limiting and did not clearly convey the fact that it would cover a full spectrum of space activities—not just the space vehicle or bus. The integrated payload and its various technical components (electronics, sensors, materials, etc.), for example, represented other critical space technologies, in addition to the space vehicle, that required advancement and development. Anderson fully realized that other directorates conducted space work connected to their technical specialty. She also suspected that other directorates were reluctant to give up their space work for fear of weakening their operations and position within the new lab. Since space was destined to become the predominant element in the Air Force's future mission, no director would willingly give up funding and resources that would prevent him or her from participating in such an important movement. If every aspect of space work transferred to one directorate, then that directorate might easily gain dominance in terms of funding, people, programs, and facilities.[25]

Even though Anderson preferred naming the new directorate *Space* or *Space Technology* instead of *Space Vehicles,* in the end she would lose this battle. Many of her colleagues in the task group believed that air and space would become the two flagship directorates in the new lab, supported by other directorates. This situation demanded a certain degree of organizational consistency and parallelism (i.e., air vehicles and space vehicles). General Paul agreed, deciding to keep the name *Space Vehicles Directorate.* In the spirit of corporate cooperation and in order to press on with the bigger and more immediate issue of getting the new lab up and running as soon as possible, Anderson accepted the decision. As Barthelemy pointed out, she was a prime example of the team player who put her personal agenda aside in favor of doing what the group thought best for the new laboratory.[26]

An even more controversial issue at the tech-directorate level focused on what to do with sensors/electronics under the single-lab organization. The old lab system scattered sensors/electronics work over three different sites—Rome, Hanscom, and Wright-Patterson. At Rome, this work came under the overall category of surveillance (various ground radars, such as phased array, laser, electro-optical, over-the-horizon, etc.) and took place in three directorates: Electronics and Reliability, Intelligence and Reconnaissance, and Surveillance and Photonics. Rome also had a contingent of scientists, permanently stationed at Hanscom and carried on Rome's unit-manning document, who worked in electromagnetics, particularly antenna and optoelectronic technologies as well as electromagnetic scattering and materials. The third major component of sensors/electronics work resided at Wright Lab's Avionics Directorate, which emphasized the integration of advanced electronics and sensors into air vehicles for the purpose of conducting aerial surveillance and detection missions. With three organizations

171

working on similar technologies, the task group had to address the best way to realign all these core technologies into a more streamlined and synergetic organization.[27]

Of the three geographically separated organizations, the Avionics Laboratory at Wright-Patterson garnered the biggest share of sensors/electronics work. At Rome, a large segment of work concentrated on information technologies associated with command, control, communications, and intelligence (C[3]I) issues. The task group initially proposed combining all sensors/electronics work with the information work at Rome. The problem with this solution, according to Barthelemy, was that "you would create a gigantic directorate . . . and no one wanted one that big." He and the others estimated that a sensors/information directorate could include as many as fourteen hundred people—a size contrary to General Paul's guidance of keeping the directorates roughly the same size or somewhere in the neighborhood of three hundred to seven hundred people.[28]

Since sensors/electronics involved so many complex technologies and integrated concepts, selecting a directorate to house them became very difficult. Barthelemy personally headed a subgroup consisting of three representatives from Rome (led by Dr. Don Bodnar) and three people from Avionics at Wright-Patterson whose job was to discuss and assess sensors/electronics work and then make a recommendation on how to organize it in the new lab. After numerous meetings, telephone calls, and E-mails, members of the subgroup presented two options to the task group. While retaining the option of merging all sensors/electronics work with information at Rome, they preferred to form two new directorates. One would deal exclusively with sensors/electronics, and the other with information technologies. A final decision on which way to go with sensors and information would not take place until after the Corporate Board's second meeting in late April.[29]

Another major reorganizational issue entailed determining what to do with Armstrong Laboratory, much of whose work dealt with medical community matters such as devising systems to protect and sustain crew members who worked with various weapon systems. Armstrong also assessed and managed health risks and ecological hazards to the Air Force war fighter. Col Ron Hill, Barthelemy's deputy, took charge of a third subgroup to

develop options for reorganizing Armstrong Lab's medical responsibilities and duties. After much debate, the subgroup recommended forming two new directorates: the Crew Effectiveness
Directorate and the Environment, Safety, and Health Directorate. As was the case with the propulsion and sensors/electron-
ics subgroups, General Paul would not announce his final decision on this recommendation until after the Corporate Board's
second meeting in April.[30]

The two Corporate Board meetings—the first one on 20–22
March and the second on 24 April—to assess the task group's
options were very important in terms of helping General Paul
and his senior leadership determine the right mixture of new
technology directorates. In preparation for each meeting,
Barthelemy and his group put together an issue paper that
described their best options and recommendations for forming
new directorates. Everyone realized that the March presentation
did not represent the group's final recommendation but a progress report that gave the Corporate Board an opportunity to
provide feedback and direction. At the first meeting, Barthelemy
told his audience that when he took the job, General Paul and
Vince Russo warned him that defining and establishing the tech
directorates would be the most difficult problem facing the task
group. He noted, however, that the tasking turned out to be
easier than expected because, as they worked the problem, it
very quickly became apparent that a natural sorting-out process
made the selection of the majority (80 percent) of tech directorates fairly straightforward.[31]

The Corporate Board reacted very favorably to Barthelemy's
briefing, agreeing to proceed with most of the group's recommendations for the tech directorates (table 9). Although no
one questioned the need for the Space Vehicles and Atmos-
pheric Vehicles (later renamed Air Vehicles) directorates under
the single lab, the Corporate Board would have to decide
whether or not propulsion work should become part of those
directorates. As mentioned above, Barthelemy recommended
keeping propulsion as a separate directorate but also presented
the alternative of merging propulsion with Space Vehicles and
Air Vehicles.[32]

As regards information and sensors/electronics technologies, Barthelemy advocated separate Information and Sensors

Table 9

Decision Options,
AFRL Directorates

• Atmospheric [later Air] Vehicles (A) two options	• Materials (M) two options
• Space Vehicles (S) two options	• Directed Energy (D)
	• Munitions (N)
• Propulsion (P)	• Crew Effectiveness (C)
• Information (I) two options	• Environment, Safety, and Health (E)
• Sensors (R)	

Source: Dr. Robert R. Barthelemy, "Report of the Technology Directorates Task Group to the Science and Technology Corporate Board," 21 March 1997.

directorates. Although he presented the Corporate Board the option of combining these two core technologies into one directorate, he advised against this because it would contain too many personnel, as mentioned previously. Despite some discussion to include electronic devices in the Munitions Directorate, that never became a major issue of contention.[33]

As for the remaining proposals, directed energy and munitions work would definitely be set up in separate directorates. Although Barthelemy's team had considered combining parts of munitions with directed energy, they dismissed that idea because of the fundamental differences between munitions (conventional weapons) and directed-energy devices such as high-power lasers and microwaves. Barthelemy also recommended a Crew Effectiveness Directorate and an Environment, Safety, and Health Directorate, to be carved out of Armstrong Lab, but General Paul was very reluctant to create them, knowing that significant personnel reductions could occur at Brooks AFB over the next few years if the labs had to deal with more manpower and/or budget cuts.[34]

The outcome of the Corporate Board's first meeting represented an important milestone in creating the new lab. Firstly, it gave both General Paul and the board a hefty dose of confidence after listening to the recommendations of Barthelemy and his team. Although Paul made no final decisions at the meeting, he was convinced that the task group was heading in the right direction and had shown significant progress after only a month of assessing the new lab's organizational structure. Secondly, Paul gave Barthelemy and his team some clear guidance on what to do next, instructing them to devote all their time and energy to the unresolved "tough issues" regarding the new directorates. This meant settling on and refining a final organizational scheme in three primary areas: propulsion, information and sensors, and health sciences. The group would brief these matters, along with recommendations about the other proposed directorates, at the next Corporate Board meeting in April.[35]

A third outcome of the March meeting was that, for all practical purposes, Paul and the Corporate Board endorsed the formation of Air Vehicles, Space Vehicles, Munitions, Directed Energy, and Materials as stand-alone directorates. Although not briefed at the meeting, AFOSR already had been identified as one of the new directorates, bringing the total to six. At the second Corporate Board meeting, Barthelemy and his team would give their best and final recommendation for the remaining directorates.[36]

The second meeting proved just as successful as the first. Between the two meetings, Barthelemy's task group had come up with its final recommendations for all the tech directorates. In addressing the various options for the three tech areas not yet fully resolved, they even went as far as drafting a spreadsheet identifying a first cut of the divisions and number of people assigned that they thought would best fit into each option. The numbers in their initial draft clearly showed that combining information and sensors would result in a large directorate consisting of nearly eight hundred people. Therefore, they supported covering these technologies by forming two more manageable directorates of approximately four hundred people each.[37]

Their numbers also revealed that Air Vehicles with propulsion would total 704 employees and Space Vehicles with propulsion would total 635. Without propulsion, the totals were 434 and 460, respectively. So to maintain an equitable balance in terms of the size of the directorates, it made more sense to Barthelemy's team to recommend an independent Propulsion Directorate of 445 people. Of course, numbers were not the only reason for establishing a Propulsion Directorate. Barthelemy had not changed his mind since his tiebreaker vote prior to the first Corporate Board meeting. He remained firm in his stance that propulsion was a self-contained and distinctive technology that deserved its own directorate. A strong "technical synergy" existed among aeropropulsion, rocket propulsion, and advanced propulsion (e.g., ramjets, scramjets, and combined-cycle engines) that a Propulsion Directorate could exploit. Barthelemy also favored a Human Effectiveness Directorate acting as an umbrella organization to take care of all the health sciences/crew-effectiveness requirements.[38]

As regards Barthelemy's recommendation of separate directorates for human effectiveness and health sciences activities, the former (e.g., man-machine interface, cockpit design, crew protection, etc.) was funded by the Air Force's S&T budget (referred to as "Program 6" funds by the budgeteers). But many of the health sciences activities (drug testing, occupational-health risk assessments, administration of environmental safety standards, clinical consulting services, etc.) were funded by the Air Force's surgeon general, using non-S&T funds (referred to as "Program 8" funds). To Paul and others, these Program 8–funded functions, although important to the Air Force, were not true S&T activities. Yet, over years of organizational restructuring, they had become gradually fused and collectively folded into the Armstrong Lab when it was formed in 1990.[39]

Paul saw the creation of AFRL as an opportunity to restrict the lab's mission solely to S&T activities. Accordingly, he proposed that the Program 8–funded activities (along with the approximately five hundred lab positions supporting that work) be transferred out of the lab structure to the Human Systems Center (HSC) at Brooks AFB. Such a transfer would require no physical movement of people. Thus, the technical synergy between Program 6 and Program 8 activities at Brooks AFB could be maintained, but HSC would now be accountable for the planning and execution of the Program 8 (non-S&T) work. After a trip by Paul to visit Lt Gen Chip Roadman, the Air Force's surgeon general, at his Bolling AFB headquarters in Washington, D.C., and after extensive discussions with the surgeon general's staff, Roadman agreed with Paul's proposal. The two generals also decided to transfer Dr. David Erwin, an SES at Armstrong Lab who was very familiar with the Program 8 work, to HSC to ensure continuity. Thus, General Paul solved the dilemma of two human-related tech directorates in AFRL—only a single Human Effectiveness Directorate would become part of the new lab.[40]

In evaluating all the pros and cons concerning the new tech directorates, General Paul kept uppermost in his thoughts one guiding principle that he wanted to implement. In his mind, one of the most important justifications for creating a single lab was to eliminate the technology seams scattered among a number of different organizations within the existing lab structure. Too many organizational boundaries fenced off similar technologies and prevented them from coming under the central control of one person who could efficiently manage and coordinate all the pieces of similar technologies as a whole. Paul had always been firmly committed to changing that situation.[41]

For example, he questioned the logic of splitting propulsion work between Edwards and Wright-Patterson: "They weren't duplicating—they weren't doing the same work—but they were both doing work related to propulsion. To really have a more coherent program, all that work ought to be under one director. That will allow one person to plan where we are going to invest our resources for all propulsion technology, no matter where it is performed." He felt the same way about sensors, information, and human effectiveness, as well as other Air Force technologies. So it made very good sense to him, in terms of effectiveness and efficiency, to support the task group's recommendations to consolidate similar technologies under one directorate whenever possible. Thus, he favored the option to create single directorates for propulsion, sensors, information, and human effectiveness.[42]

After weighing all the recommendations presented at the meeting of 24 April and after consulting with his senior staff, General Paul made the final decision, identifying the tech directorates that would make up the new laboratory. As it turned out, his task did not prove too difficult because of the completeness and high quality of the work done in Barthelemy's task group. Ten new directorates appealed to Paul because that number confirmed his initial "gut feeling" that eight to 12 new directorates seemed about right. In short, Paul accepted and approved the recommendations of the task group. Paul then made one of his top priorities the dissemination of this critical information to everyone in the organization as a way of dispelling any rumors and sharing the steady progress toward the stand-up of the single lab. During the second week of May, he posted a message on his web site, naming the new tech directorates (fig. 16).[43] On 31 July, Paul posted another message that explained his decision:

> By consolidating our current 22 technology directorates to nine, our new technology directorates will be large enough to have critical mass and to exercise the flexibility needed to realign themselves internally to meet the technical challenges of the time. These directorates are the "engine" of AFRL and represent the Air Force's core technology areas for the future. Additionally, our hope is that nine "boxes" with their attendant smaller number of divisions and branches will have less overhead manpower than 22 "boxes," thus contributing to a more efficient overall organization as we continue to take manpower reductions

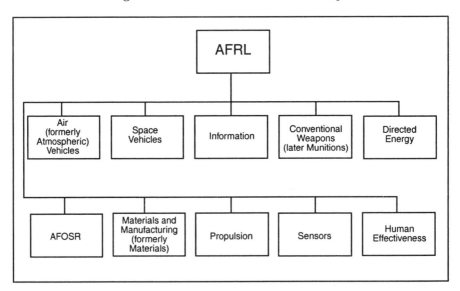

Figure 16. AFRL Technology Directorates (From message, General Paul's web site, subject: AFRL Progress Report—Directorates and More; on-line, Internet, 14 May 1997, available from http://stbbs.wpafb.af.mil/STBBS/labs/single-lab/updates.htm)

in compliance with previous mandates. Grouped with our nine technology directorates is AFOSR, which has remained largely intact and will continue to manage our basic research program.[44]

With the establishment of the new tech directorates, it became clear that similar technologies, formerly dispersed throughout many directorates in the four labs, now became consolidated under one of the nine new directorates. This change represented a major step forward because it conformed to the long-held premise that the Air Force laboratory system needed to do a better job of eliminating "fragmented technologies." Paul described this state of affairs as " 'islands of technology'—islands with rickety or nonexistent bridges between them." The new organizational structure went a long way toward solving this problem by organizing along well-defined technology disciplines (fig. 17). For example, under the old lab system, sensor

	Wright Lab						Rome Lab				Phillips Lab						Armstrong Lab					
	AA	PO	FI	ML	MT	MN	C³	ER	IR	OC	GP	LI	RK	SX	VT	WS	AO	CF	EQ	HR	OE	PS
Air Vehicles			X																			
Space Vehicles											X	X		X	X	X						
✓ Information	X						X	X	X	X												
Munitions					X																	
Directed Energy												X			X							
✓ Materials/Manufacturing			X	X	X																X	
✓ Sensors	X						X	X	X													
✓ Propulsion		X											X									
✓ Human Effectiveness	X		X														X	X		X	X	HSC

✓ = Major cross-lab synergy opportunities

Wright Lab Directorates:
AA= Avionics
PO= Aero Propulsion and Power
FI= Flight Dynamics
ML= Materials
MT= Manufacturing Technology
MN= Munitions

Phillips Lab Directorates:
GP= Geophysics
LI= Lasers and Imaging
RK= Rocket Propulsion
SX= Space Experiments
VT= Space Technology
WS= Advanced Weapons

Rome Lab Directorates:
C³= Command, Control, and Communications
ER= Electronics and Reliability
IR= Intelligence and Reconnaissance
OC= Surveillance and Photonics

Armstrong Lab Directorates:
AO= Aerospace Medicine
CF= Crew Systems
EQ= Environics
HR= Human Resources
OE= Occupational and Environmental Health
PS= Office of Preventive Health Service Agency (transferred out of lab to Human Systems Center [HSC] on formation of the single lab)

Figure 17. AFRL Directorate Matrix (From Maj Gen Richard R. Paul to author, E-mail, subject: Conversion Matrix of Four Lab Directorates to AFRL Directorates, 2 February 2000, with attached chart "AFRL Directorate Matrix")

work took place at two labs (Rome and Wright) and was spread among four different directorates (Avionics, Electromagnetics and Reliability, Intelligence and Reconnaissance, and Surveillance and Photonics). Now, under the single-lab concept of operation, all sensors work transferred to the new Sensors Directorate. Similarly, aerospace medicine and occupational health studies, previously assigned to six directorates at the Wright and Armstrong labs, became centralized under one new Human Effectiveness Directorate, set up at Wright-Patterson. Viccellio and Paul had wanted to accomplish these kinds of efficiency reforms since the earliest planning stages of the laboratory-restructuring effort.[45]

Consolidating similar technologies into a smaller number of new tech directorates altered the geographic profile of AFRL, which extended from coast to coast. Half of the new tech directorates resided at Wright-Patterson with the new lab headquarters, and the other five were located in New York, Washington, D.C., Florida, and New Mexico (fig. 18).[46]

Nailing down the tech directorates cleared the way for removing many obstacles and uncertainties standing in the way of planning for the stand-up of the laboratory. The next step entailed making sure the AFRL strategic plan under development spelled out what the tech directorates were expected to

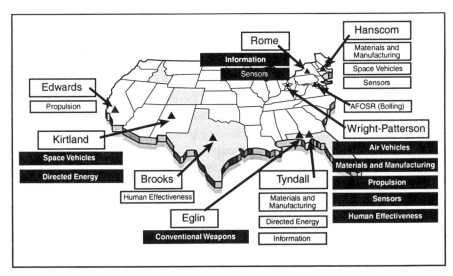

Figure 18. AFRL Locations (From briefing chart, "AFRL Locations," n.d.)

accomplish. General Paul, Vince Russo, and other senior staff members attended an off-site meeting 12–13 May to begin a dialogue for building a single-lab strategic plan. The first version of the strategic plan, prepared at the end of May, laid out the vision and mission for the new lab and served as the basis for all the work conducted by the tech directorates. Looking to the future, the plan publicized the new lab's vision as "the best people providing the best technologies for the world's best Air Force." To make this happen, the tech directorates had to accomplish the lab's mission: "to lead the discovery, development, and timely transition of affordable, integrated technologies that keep our Air Force the best in the world."[47]

The strategic plan informed all the tech directorates of their responsibility for assuring the timely development of core technologies for the operational Air Force to enhance its future war-fighting capabilities. Specifically, according to the AFRL strategic plan, the technology directorates

> will be accountable for the overall health and welfare of their assigned resources. They will be the stewards of the 6.1 [Basic Research], 6.2 [Applied Research], 6.3 [Advanced Technology Development], and external funds assigned to and executed by them. Additionally, they will be the advocates and strategists for the technology disciplines within their respective areas of responsibility. Their leaders will be national and international spokespersons for their areas of expertise. . . . The TDs [Technical Directors] will develop integrated and balanced full spectrum budgets and programs that are responsive to the needs of the Air Force. . . . A major responsibility of the TDs will be the development, training, appraisal and recognition of their people.[48]

Paul also pointed out that the products of the tech directorates, especially in the 6.3 program areas, would require the "integration of technologies of multiple AFRL directorates." In other words, many different tech directorates would contribute their specialized technical component to an integrated product or system. Because today's highly complex systems consist of a wide spectrum of scientific disciplines contributing to the development of advanced hardware (i.e., sensors, electronics, materials, avionics, power sources, etc.), rarely would cases arise in which one tech directorate would have responsibility for the totality of any new product. So emphasis would be on exploiting technology "synergies" among directorates as the most effective and cost-effective way of doing business. All this meant that the new tech directors inherited a huge challenge and responsibility to meet these ambitious goals. But their actions would ultimately determine, to a great degree, the success or failure of the new laboratory in the years to come.[49]

Disclosure of the new tech directorates by General Paul had two very immediate and important effects. Firstly, it sent an unequivocal signal to everyone that a major organizational shake-up would definitely happen, once and for all erasing any lingering doubts about preserving the old four-lab system or stopping the reorganization with phase I. There was no turning back. Secondly, the workload for forming the single lab would not diminish. One of the most pressing tasks ahead involved determining what specific divisions and branches would become organizational components of each of the new tech directorates. That inescapable assignment had to be accomplished before the lab could stand up in October.

General Paul's announcement officially ended the work of the Technology Directorate Task Group. However, he elected to keep Dr. Barthelemy and Colonel Hill on board for several more months because of the great amount of unfinished business, particularly in the area of personnel. Someone needed to identify and review all civilian and military positions in the old lab to determine slot assignments in each of the new tech directorates on AFRL's new unit-manning document. Paul counted on Barthelemy and Hill to tackle this extremely detailed and labor-intensive process of looking at over six thousand positions; he had to make sure that every slot transferred to the right directorate, division, or branch to ensure that no organization lost positions in the process and that everyone kept his or her job. Although everyone would still have a job in October, review of the current unit-manning documents included the identification of vacancies and filled positions tagged for possible elimination as part of phase IIB's downsizing plan. Barthelemy and Hill were well suited for this huge job, having gained valuable hands-on experience and having developed personal contacts labwide while working on the task group. They would depend on those contacts to help them work the difficult unit-manning-document issues.[50]

Before the assignment of specific positions, new divisions and branches had to be defined and established in each of the new directorates. Some of the initial proposals for the organization of divisions and branches had already emerged as part of the options briefed at the second Corporate Board meeting. To keep this process moving forward, Barthelemy and Hill worked closely with the current lab commanders and their staffs, tech directors, division and branch chiefs, personnel officials, and others. They would continue this work with the new tech directors, selected in the summer of 1997.[51]

Selection of the tech directors was extremely important because those individuals would lead an untested laboratory organization into the next century. This called for a fair-minded selection process to ensure that only the most qualified people ended up in the most influential and powerful positions in the lab. A strong proponent of shared leadership, General Paul wanted the proportion of military and civilian tech directors to reflect the composition of the lab's workforce. The current leadership consisted of three military commanders and one civilian director, despite the fact that civilian employees outnumbered military by roughly three to one. Paul realized that with the movement to a single lab, he had a once-in-a-lifetime opportunity to readjust the ratio of military and civilians occupying key leadership positions.[52]

Placing more civilians in positions of high authority traditionally reserved for senior-grade officers seemed a radical departure from the current lab-management system. In fact, a number of colonels in the four labs viewed this change as the first round of reducing the military presence in the labs. They worried that this would reduce the chances of senior officers in the science and engineering career fields to realistically compete for promotion to general officer. But General Paul did not see it that way, believing that the new lab would support about the same number of full colonels (and other military grades) as the old lab structure. Although the overall number of military and civilians had declined as a result of recent personnel cutbacks, the proportion of the military/civilian mix had remained the same. Therefore, he favored moving in a new direction, anticipating a gradual policy shift toward placing a higher percentage of civilians in key leadership positions. For example, AFMC had implemented a plan to replace one of the three existing lab military commanders with a civilian SES, which would change the ratio to 50/50. Although the new ratio still would not reflect the true workforce population, at least it would be a step in the right direction.[53]

Paul had to wrestle with another problem associated with placing military members in positions of high authority in the lab. If he appointed a colonel as a tech director, then his long-term plan called for appointing an SES as the deputy of that directorate. However, the Air Force Executive Resources Board (ERB), the authority for approving all SES assignments, had a policy stating that an SES could not work for a colonel because it considered an SES the equivalent of a general officer. The only way around this prohibition entailed requesting an exception from the board. Indeed, when Paul commanded Wright Laboratory as a colonel from July 1988 to July 1992, he had several SESes reporting to him. General Paul planned to request exceptions to ERB's policy on a case-by-case basis, but until then, he had to proceed with the assignment of key leadership positions. Thus, until otherwise notified, he would have to appoint a GS-15 rather than an SES to serve as the deputy of any tech directorate led by a colonel.[54]

Another factor complicating the filling of the new positions was the availability of only 10 director slots. If three were reserved for senior military officers, then only seven were open to civilians. Since 22 individuals had SES jobs—mostly tech directors and a few senior staff slots—what would become of the ones not selected to fill the director vacancies? Some of them might need to move down a step in the organizational chain to become division chiefs while retaining their SES ranking—with an attendant loss of prestige and authority. The long-range plan envisioned that over the next few years some SESes would retire or move to other jobs in other organizations, after which their positions would be dropped from the unit-manning document. Eventually, under the new lab, SES jobs would be assigned only to tech directors, the AFRL executive director, and about five other key staff positions, including the director of XP and the chief scientist.[55]

As a first step, General Paul instructed Dr. Daniel and Dr. Barthelemy to prepare separate lists of possible candidates for the tech director positions. Paul solicited Barthelemy's input because, as leader of the task group, he had worked with, observed, and judged most of the tech leaders in the lab almost daily over the last three months. In developing his list, Barthelemy could not consult with any member of the task group. The process—a sensitive matter involving personalities and career progression—would remain private and not subject to democratic vote, with input restricted to only a few senior people designated by General Paul. The general knew that the final decision would be his—and his alone—as commander of the new lab. Although Barthelemy received no comment on his nominations, 90 percent of Paul's final selections were names on his list.[56]

Despite Barthelemy's contribution, Dr. Daniel, deputy director of AFMC/ST, played a more active and influential role in terms of working directly with General Paul on the selection process. Because of the sensitive political nature of this exercise, for the most part, Daniel worked alone and sought little advice in preparing his list. He did, however, consult with Colonel Markisello from time to time and talked with a few prospective candidates one-on-one, as needed. Although purposely treating the list as a private, close-hold item, Daniel remained open-minded in evaluating candidates: "I took all the SESes and said I am going to go through this with an unconstrained attitude. I would write down every position where I thought Sally, Joe, or Larry could go. And if I had to write Sally down 14 times, I would write down 14 times every place where I thought Sally, Joe, or Larry could be placed. Then I went back, looked at the jobs, and made a list of the top three who could do that job."[57]

After making a first cut, Daniel met with General Paul on numerous occasions to discuss the technical and leadership qualifications of each candidate, investing a significant amount of time going over the top three candidates for each position. To get a better feel for a particular SES's future plans, Daniel would meet with that person. If, for instance, he learned that the candidate planned to retire in the near future, then he plugged that piece of information into the final selection process. If some were leading candidates for more than one tech director job, Daniel and Paul discussed their qualifications at length, wanting to make sure they selected the right person who would best serve the entire laboratory.[58]

Finalizing the selection of the new tech directors turned out to be a very time-consuming procedure, primarily because the appointment of each SES tech director required approval from the AFMC commander and the ERB. According to Dr. Daniel, "SES assignments have to go all the way to the undersecretary of the Air Force. It is a very formal process. We don't have the authority. Gen [George T.] Babbitt [AFMC commander] doesn't have the authority to reassign SESes. He can propose . . . but can't give the final approval. So we put a list together, briefed it to General Paul and got his buy-in—briefed it to General Babbitt and got his buy-in. Then I personally took it to the Air Force Executive Administration Board, briefed Darleen Druyun—got her buy-in, and the undersecretary approved it."[59]

During the second week of June, General Paul posted a message on his web site informing people of the progress of the selection of the new lab leaders and explaining that he was in the "final phases of proposing a set of provisional leaders to AFMC/CC." The list consisted of his choices for tech directors and the two key staff directors (i.e., XP and the DS directorates). Paul planned to release the names of all the "provisional" directors once General Babbitt approved the list. Final approval would not come until after the ERB gave its blessing concerning assignment locations. In the meantime, Paul became somewhat impatient, not wanting to prolong passing on to everyone the identity of the new directors, which he considered vital information. The sooner the better was Paul's philosophy because of the number of practical considerations at stake. Most importantly, he wanted the new leaders identified so they could "begin working the details of the organizations, which they will end up leading." Paul also was anxious to get things moving since the scheduled AFRL stand-up was less than four months away. During that time, he wanted the new directors working vigorously with people at all levels to solve some very difficult and labor-intensive organizational problems—particularly, determining the number of divisions and branches and the assignments of each individual in the new laboratory.[60]

Although Paul had hoped to name all the new tech directors in one announcement, that never transpired because of the inevitable tangle of rigid procedural rules and regulations imposed by a government system that seemed to move any personnel action at a snail's pace. However, timeliness was an extremely valuable commodity to Paul and the more than five hundred other individuals working the formation of the new lab. Consequently, in August (by that time, General Babbitt had given his approval, but ERB had not issued final approval of *all* the nominees) he could not wait any longer and decided to share the names of all the "approved" tech directors. He would announce the other directors—the ones he labelled "to be determined" (TBD)—after the personnel system had approved them.[61]

In the third week of August, less than two months before the planned stand-up, Paul announced the names of the eight SESes and two colonels he had selected as new tech directors (table 10). Although this proportion did not reflect

Table 10

First Directors, Deputies, and Chief Scientists of AFRL

Directorate/Symbol	Director	Deputy Director	Chief Scientist
Air Vehicles/VA	TBD (colonel)	Mr. Terry Neighbor (DR-4)	Dr. Don Paul (ST)
Space Vehicles/VS	Ms. Christine Anderson (SES)	Col Bruce Thieman	Dr. Janet Fender (ST)
Information/IF	Mr. Ray Urtz (SES)	TBD (colonel)	Mr. John Granier (ST)
Munitions/MN	Col Bob Wood	Mr. Steve Korn (DR-4)	Dr. Bob Sierakowski (ST-IPA)
Directed Energy/DE	Dr. Earl Good (SES)	TBD (colonel)	Dr. Barry Hogge (ST)
Materials and Manufacturing/ML	Dr. Vince Russo (SES)	Col Don Kitchen	Dr. Wade Adams (ST)
Sensors/SN	Mr. Les McFawn (SES)	Col Gerry O'Conner	Dr. Don Bodnar (ST-IPA)
Propulsion/PR	Dr. Tom Curran (SES)	Col John Rogacki	Dr. Alan Garscadden (ST)
Human Effectiveness/HE	Mr. Jim Brinkley (SES)	TBD (colonel)	Dr. Ken Boff (ST)
AFOSR	Dr. Joe Janni (SES)	Col Bob Herklotz	N/A

Source: Message, General Paul's web site, subject: Key Leaders for the AFRL Phase II Organization; on-line, Internet, 22 August 1997, available from http://stbbs.wpafb.af.mil/STBBS/labs/single-lab/updates.htm.

Ms. Christine Anderson
(Space Vehicles)

Mr. Terry Neighbor
(Air Vehicles [acting])

Col Bob Wood
(Munitions)

Dr. Vince Russo
(Materials and Manufacturing)

Dr. Earl Good
(Directed Energy)

Dr. Joseph Janni
(AFOSR)

Dr. Tom Curran
(Propulsion)

Mr. Les McFawn
(Sensors)

Mr. Jim Brinkley
(Human Effectiveness)

Mr. Ray Urtz
(Information)

AFRL's First Technology Directors as of October 1997

the desired 70/30 percent civilian/military mix, Paul intended to achieve that ratio shortly after stand-up (this occurred in the spring of 1998, when Dr. Tom Curran retired and Colonel Rogacki took over as the new Propulsion director and when Col Bob Wood moved from director of Munitions to director of Air Vehicles, Col Gerry Daugherty replacing him as Munitions director).[62]

Since General Paul's announcement placed only eight SESes in key jobs as tech directors, he released the names of 12 other SESes he had assigned to top-level positions, three of whom became associate directors to Space Vehicles, Materials and Manufacturing, and Sensors. His long-range plan called for keeping these slots until a person went to another job or retired, at which time the SES position would go away. For instance, when Dr. Hal Roth, associate director of Space Vehicles, retired a year after the stand-up of the lab, General Paul did not hire a new SES to take his place. Four other SESes who headed up four basic research-science disciplines at AFOSR brought the total to 15 SESes working in the 10 tech directorates.[63]

The remaining six SESes belonged to Paul's command section and central staff. Appointed as the new lab's executive director, Dr. Daniel retained his SES position in the command section. The lab's chief scientist slot was also reserved for an SES. At the senior-staff level, the directors of XP; Corporate Information; and the Washington, D.C., office, as well as the associate director for investment strategy all became SES positions. So by August, General Paul had succeeded in justifying and placing 21 of 22 SESes in permanent positions in the new laboratory structure. These strong-minded and experienced individuals would assume responsibility for leading a major thrust over the next two months to prepare for the stand-up of the new lab. In the end, their efforts resulted in radically changing the cumbersome organizational configuration of the old four-lab structure to a more streamlined and efficient single laboratory. They did so by consolidating similar or related technologies into fewer labs, directorates, divisions, branches, and planning staffs (table 11).[64]

Table 11

Air Force Research Laboratory: Before and After

Before	Organization Element	After
4	Labs	1
22	Directorates	9
115	Divisions	50
351	Branches	192
5	Planning Staffs	1
	AFRL is more efficient *and* more focused.	

Source: Briefing, Maj Gen Richard R. Paul, subject: Air Force Research Laboratory: Overview Briefing to Industry, February 1998.

Notes

1. Maj Gen Richard R. Paul, interviewed by author, 2 March 1998 and 22 November 1999; and Dr. Robert R. Barthelemy, interviewed by author, 28 January 2000.

2. Dr. Vince Russo, interviewed by author, 4 February 1998.

3. Ibid.; and Paul interview, 2 March 1998.

4. Paul interview, 2 March 1998; and USAF biography, "Dr. Robert R. Barthelemy," Aeronautical Systems Center Public Affairs Office, Wright-Patterson AFB, Ohio, March 1993.

5. Paul interview, 2 March 1998; and Russo interview.

6. Dr. Robert R. Barthelemy, interviewed by author, 6 February 1998; and message, General Paul's web site, subject: Midcourse Heading Check; on-line, Internet, 24 March 1997, available from http://stbbs.wpafb.af.mil/STBBS/labs/single-lab/updates.htm.

7. Barthelemy interview, 6 February 1998; and Paul message, 24 March 1997.

8. Barthelemy interview, 6 February 1998; Paul message 24 March 1997; and Ms. Christine Anderson, director, Space Vehicles Directorate, interviewed by author, 18 March 1998.

9. Barthelemy interview, 6 February 1998; Paul message, 24 March 1997; Anderson interview; and Paul interview, 2 March 1998.

10. Dr. Robert R. Barthelemy, "Report of the Technology Directorates Task Group to the Science and Technology Corporate Board," 21 March 1997.

11. Dr. Robert R. Barthelemy to Technology Directorate Committee members, letter, subject: Technology Group Meeting, 20 February 1997; and Col Ron Hill to Technology Group members, E-mail, subject: Technology Directorate Committee Members, 19 February 1997.

12. Barthelemy interview, 6 February 1998; and message, General Paul's web site, subject: 20–22 March Corporate Board Meeting; on-line, Internet, 20 March 1997, available from http://stbbs.wpafb.af.mil/STBBS/labs/single-lab/updates.htm.

13. Barthelemy interview, 6 February 1998.

14. Ibid.; and Paul interview, 22 November 1999.

15. Barthelemy interview, 6 February 1998; and Anderson interview.

16. Barthelemy interview, 6 February 1998; and Anderson interview.

17. Barthelemy, "Report of the Technology Directorates Task Group"; and Maj Gen Richard R. Paul to SMC/CC, letter, subject: Single Air Force Laboratory, 12 February 1997.

18. Barthelemy, "Report of the Technology Directorates Task Group"; Paul letter; Barthelemy interviews, 6 February 1998 and 28 January 2000; and briefing, Maj Gen Richard R. Paul, subject: Air Force Research Laboratory: Heading Check, 29 April 1997.

19. Barthelemy, "Report of the Technology Directorates Task Group"; Paul letter; Barthelemy interviews, 6 February 1998 and 28 January 2000; and Paul briefing.

20. Barthelemy, "Report of the Technology Directorates Task Group"; Paul letter; Barthelemy interviews, 6 February 1998 and 28 January 2000; and Paul briefing

21. Barthelemy, "Report of the Technology Directorates Task Group"; Paul letter; Barthelemy interviews, 6 February 1998 and 28 January 2000; Paul briefing; Col John Rogacki, interviewed by author, 9 June 1998; Dr. Tom Curran, interviewed by author, 4 March 1998; Anderson interview; Edward T. Curran, director, Aero Propulsion and Power Directorate, memorandum to Maj Gen Richard R. Paul, subject: A Status Report, 5 June 1997; and Fred Oliver, WL/POM, memorandum to Dr. Garscadden, WL/CA, subject: Your Friday Briefing, 3 April 1997, with attached briefing, "AFRL's Aerospace Propulsion and Power Directorate."

22. Anderson interview; Dr. Keith Richey, interviewed by author, 5 February 1998; and Rogacki interview.

23. Anderson interview; Barthelemy interview, 28 January 2000; and Joe Sciabica, memorandum to Col Ron Hill, subject: Space Flight Directorate, 17 March 1997.

24. Anderson interview; Barthelemy interview, 28 January 2000; and Sciabica memorandum.

25. Anderson interview.

26. Barthelemy interview, 6 February 1998; and Maj Gen Richard R. Paul to author, E-mail, subject: Answers to Your Questions, 23 March 1998.

27. Barthelemy interview, 6 February 1998; Paul E-mail; Les McFawn, interviewed by author, 30 July 1998; and Les McFawn, briefing chart, "Sensors Directorate," 15 July 1998.

28. Barthelemy interviews, 6 February 1998 and 28 January 2000; Barthelemy, "Report of the Technology Directorates Task Group"; message,

General Paul's web site, subject: AFRL Progress Report—Directorates and More; on-line, Internet, 14 May 1997, available from http://stbbs.wpafb.af.mil/STBBS/labs/single-lab/updates.htm; and Paul briefing.

29. Barthelemy interviews, 6 February 1998 and 28 January 2000; Paul messages, 20 and 24 March 1997; message, General Paul's web site, subject: AFRL Phase II: A Decision Process Flowchart; on-line, Internet, 25 April 1997, available from http://stbbs.wpafb.af.mil/STBBS/labs/single-lab/updates.htm; and Don Bodnar, RL/CA, "Report of the Subcommittee on Reducing Overlaps between Rome Laboratory and Wright Laboratory/Avionics Directorate," 13 March 1997.

30. Barthelemy, "Report of the Technology Directorates Task Group"; and Barthelemy interview, 28 January 2000.

31. Barthelemy interviews, 6 February 1998 and 28 January 2000; Paul messages, 20 March and 25 April 1997.

32. Barthelemy interviews, 6 February 1998 and 28 January 2000; Paul messages, 20 March and 25 April 1997; and Barthelemy, "Report of the Technology Directorates Task Group."

33. Barthelemy, "Report of the Technology Directorates Task Group"; Barthelemy interviews, 6 February 1998 and 28 January 2000; Russo interview; and Paul briefing.

34. Barthelemy, "Report of the Technology Directorates Task Group"; Barthelemy interviews, 6 February 1998 and 28 January 2000; Russo interview; and Paul briefing.

35. Barthelemy, "Report of the Technology Directorates Task Group"; Barthelemy interviews, 6 February 1998 and 28 January 2000; Russo interview; Paul briefing; Paul message, 24 March 1997; and Maj Gen Richard R. Paul to author, E-mail, subject: AFRL History Project, 16 March 1998.

36. Barthelemy, "Report of the Technology Directorates Task Group"; Barthelemy interviews, 6 February 1998 and 28 January 2000; Russo interview; Paul briefing; Paul messages, 24 March and 25 April 1997; and Paul E-mail, 16 March 1998.

37. Dr. Robert R. Barthelemy, spreadsheet, "AFRL Structure," 23 April 1997; Barthelemy interview, 28 January 2000; and McFawn interview.

38. Barthelemy spreadsheet; and Barthelemy interviews, 6 February 1998 and 28 January 2000.

39. Maj Gen Richard R. Paul, notes, subject: Lab Reorganization, 20 April 2000.

40. Ibid.

41. Paul interview.

42. Ibid.

43. Paul message, 14 May 1997; message, General Paul's web site, subject: AFRL Phase II Status; on-line, Internet, 31 July 1997, available from http://stbbs.wpafb.af.mil/STBBS/labs/single-lab/updates.htm; Barthelemy, "Report of the Technology Directorates Task Group"; Paul interviews, 2 March 1998 and 22 November 1999; and Paul briefing.

44. Paul message, 31 July 1997.

45. Maj Gen Richard R. Paul to author, E-mail, subject: Conversion Matrix of Four Lab Directorates to AFRL Directorates, 2 February 2000; message, General Paul's web site, subject: The AFRL Corporate Information Office; on-line, Internet, 7 May 1997, available from http://stbbs.wpafb.af.mil/STBBS/labs/single-lab/updates.htm; and Paul briefing.

46. Briefing chart, "AFRL Locations," n.d.

47. Dr. Vince Russo, memorandum to AFRL Corporate Board, subject: Charts from 12–13 May Off-Site, 27 May 1997, with attached Air Force Research Laboratory Strategic Plan, May 1997.

48. Dr. Vince Russo, working paper, "Air Force Research Laboratory: Strategic Objectives" [May 1997].

49. Russo memorandum; and Paul message, 14 May 1997.

50. Barthelemy interview, 28 January 2000.

51. Ibid.; and message, General Paul's web site, subject: Key Leaders for the AFRL Phase II Organization; on-line, Internet, 22 August 1997, available from http://stbbs.wpafb.af.mil/STBBS/labs/single-lab/updates.htm.

52. Vannevar Bush, science advisor to President Franklin Roosevelt during World War II, was a strong proponent of the shared-leadership concept. Over 50 years ago, he advocated assigning civilians to key leadership positions in the military services as one of the most efficient ways to advance S&T. See G. Pascal Zachary, *Endless Frontier: Vannevar Bush, Engineer of the American Century* (New York: Free Press, 1997); Paul interview, 2 March 1998; and Dr. Don Daniel, interviewed by author, 27 July 1998.

53. Colonel Rogacki, a senior military leader in the lab system, agreed with General Paul that promotion opportunities under the new lab would not be diminished. Other colonels assigned to the lab, such as Robert A. Duryea and William G. Heckathorn, disagreed. Paul interview, 2 March 1998; Rogacki interview; and Russo interview.

54. Paul interview, 2 March 1998.

55. Ibid.; Barthelemy interview, 6 February 1998; and Daniel interview.

56. Barthelemy interview, 6 February 1998; and Daniel interview.

57. Daniel interview; Col Dennis Markisello, interviewed by author, 6 February 1998; and Ms. Wendy Campbell, interviewed by author, 11 June 1998.

58. Daniel interview.

59. Ibid.

60. Message, General Paul's web site, subject: Status Report; on-line, Internet, 12 June 1997, available from http://stbbs.wpafb.af.mil/STBBS/labs/single-lab/updates.htm; and Dr. Don Daniel to Dr. Earl Good et al., E-mail, subject: AFRL Senior Executive Assignments, 23 June 1997.

61. Paul message, 12 June 1997; and Daniel E-mail.

62. Paul messages, 12 June and 22 August 1997; and Daniel E-mail.

63. Paul message, 22 August 1997; Capt Chuck Helwig, interviewed by author, 3 February 1997; and Campbell interview.

64. Paul interview, 2 March 1998; Daniel interview; and briefing, Maj Gen Richard R. Paul, subject: Air Force Research Laboratory: Overview Briefing to Industry, February 1998.

Chapter 11

Getting the Message Out

In the third week of January 1997, in compliance with AFI 38-101, *Air Force Organization*, General Paul sent an organization change request (OCR) proposing the creation of a single Air Force laboratory to Maj Gen Michael C. Kostelnik, director of plans at Headquarters AFMC. In turn, Kostelnik submitted the OCR package to the Air Staff at Headquarters Air Force on 3 February, explaining in his cover letter that the OCR represented AFMC's response to "the goals of the congressionally-mandated Vision 21 initiative" and section 277 of the National Defense Authorization Act for FY 1996 to reduce infrastructure costs. The benefits of moving to a single lab, he pointed out, included a more streamlined operation with less overhead, a reduction in duplication of similar technologies at multiple laboratory sites, and one commander accountable for all facets of the consolidated laboratory. To keep the phase I implementation on schedule, he requested that Headquarters Air Force approve the OCR by 15 March. If all went according to plan, he intended the lab to stand up by the end of March.[1]

Phase I Stand-Up of the Laboratory

A number of salient points in the OCR very specifically identified the changes necessary to establish the single lab—most obviously, the consolidation of the four current labs and AFOSR into a single lab. The name of the new lab would be the Air Force Research Laboratory, which accurately and concisely reflected the mission of the organization and fit in with the names of its counterparts in the Army and Navy—the Army Research Laboratory and the Naval Research Laboratory. The term *Research* differentiated it from the Air Force battle labs, which focused on tactics and technology insertion, as opposed to AFRL's mission of developing and demonstrating new technologies. The name *Air Force Research and Development Laboratory*, favored earlier by some parties, was discarded because AFMC's product centers—not the labs—performed the development function for weapon systems.[2]

According to the OCR, the mission of the new lab entailed providing "the science and technology required to enable the US Air Force to defend the United States through control and exploitation of air and space." This new mission statement, replacing the statements of the four labs and AFOSR, would become effective at the stand-up of the interim lab in April, which would mark the end of the phase I implementation process. Each of the 10 new tech directorates would formulate its own mission statement.[3]

At the end of March, AFMC received an approval letter from the Directorate of Manpower, Organization, and Quality at Headquarters Air Force, stating that AFRL "is constituted [and] assigned [to AFMC] effective 31 March." As an approved organization and recognized as a legitimate Air Force unit by the secretary of the Air Force, AFRL would activate "on or about" 16 April. However, General Viccellio, in keeping with his policy of pressing on with the lab consolidation, had no intention of waiting until then to set up the new lab. The day after he received notification of approval, Viccellio directed Col Jacob Kessel, chief of manpower and organization at AFMC, to issue Special Order GA-9, activating AFRL effective 8 April 1997. This occurred without delay, marking the official end of the first phase of the implementation.[4]

Very little fanfare accompanied the establishment of the new lab on 8 April because of the temporary nature of this measure, designed to keep the process on track. Over the next six months, many lab-implementation issues would require study, work, and resolution before the "end state" single lab would stand up in October. By that time, all the new tech directorates would have been identified and assigned as specific detachments to AFRL. Nevertheless, the interim stand-up remained important because it represented the first tangible evidence that the new lab would become a reality. To commemorate this event, a short, modest ceremony took place on 8 April at the AFMC Spring Commander's Conference held at Wright-Patterson, with General Viccellio reading the orders activating the single lab and unveiling the proposed new AFRL emblem (see below). General Paul then gave a short briefing to all the AFMC commanders, recounting the events leading up to the stand-up and promising a much bigger ceremony for phase II in the fall. At that time, more people would have an opportunity to participate and recognize the rich heritage of the four labs as AFRL leaders ushered in the dawn of a new S&T era anchored by the newly established laboratory.[5]

The April stand-up produced some very immediate and practical changes. Firstly, although the four labs would still exist as named units until the completion of phase II in October, the lab commanders would now report to the commander of AFRL—General Paul—instead of the commanders of the AFMC product centers (Aeronautical Systems Center, Electronic Systems Center, Human Systems Center, and Space and Missile Systems Center). The four lab commanders would continue to perform their duties as commanders and would simultaneously serve as AFRL product executives. As for General Paul, not only would he serve as AFRL commander, he would also become the Air Force's technology executive officer reporting to the service acquisition executive on investment and programmatic matters, as well as become director of Science and Technology at Headquarters AFMC. All of these leadership changes served to centralize the span of control of the new laboratory under one commander, making for a more effective and efficient organization. This was especially true as regards personnel matters, since the lab's leadership could now speak in one voice to determine the best course of action for reaching the mandatory personnel reductions planned for the next few years.[6]

Secondly, as of 8 April, the OCR approval letter instructed AFRL to transfer all personnel (a total of 7,075 military and civilian positions) from the four labs and AFOSR to a new AFRL unit-manning document. No longer would the laboratories' manpower be assigned to the product centers' manning documents. Ninety-three positions would transfer to the new document from General Paul's Science and Technology Directorate; 144 from AFOSR; 1,739 from Phillips Lab; 1,505 from Armstrong Lab; 2,547 from Wright Lab; and 1,047 from Rome Lab. Every one of these positions had to be matched against a position authorized on the new unit-manning document prior to the stand-up in October, an extremely time-consuming and taxing exercise that resembled trying to put together a 7,075-piece jigsaw puzzle. Moreover, the change in the new lab's organizational structure would prevent a one-to-one correspondence in the shifting of jobs from the old unit-manning document to the new one. Some jobs would assume more or fewer duties, entailing changes in job descriptions and placing them in appropriate slots on the new document.[7]

Thirdly, the OCR approval also involved restructuring the staff of the new lab headquarters, with the new command section consisting of the commander and executive director. Key components of the commander's staff would include the chief scientist, the four product executives, and the directors of Plans and Programs (AFRL/XP) and Operations and Support (AFRL/DS). Under this arrangement to centralize overall planning activities and establish an immediate integrated planning function to facilitate the transition to phase II, the four labs' XP offices were directed to report to AFRL/XP. The new staff organization would have to submit the phase II OCR to Headquarters Air Force at least 60 days prior to the implementation date of October 1997. As it turned out, building the phase II OCR became an all-consuming effort that did not see completion until early August.[8]

Looking back, it is rather remarkable that what started as a fragmentary strategic concept in the minds of a few people in January 1997 turned into reality a short three months later with the interim stand-up of the single laboratory. This major event symbolized a radical and fundamental change within the Air Force in terms of how it planned to conduct its S&T business in the future. To bolster its S&T mission, General Paul thought it very important to craft a distinctive emblem that would promote and identify what the laboratory stood for. Just as the "swoosh" check gives instant recognition to Nike products, a creative AFRL emblem would convey a comparable message to any audience, underscoring the importance of the S&T mission throughout the Air Force and DOD.[9]

Since the unit emblem would represent the totality of the new laboratory, Paul decided to "go to the troops" to solicit their ideas. In February 1997, he sent a letter to the workforce encouraging everyone to submit a heraldry design that would best exemplify the new lab. In a pleasantly surprising response, workers from the four labs, AFOSR, and AFMC/ST turned in 45 full-color entries. General Paul, Dr. Daniel, and Dr. Russo reviewed and evaluated each entry, narrowing the field to eight emblem designs. Paul declared that the "very good" submissions made it quite difficult for him to select a winner. But on 24 March, he opted for a synthesis of two different designs proposed by the team of Mr. Rogelio Burgos and 1st Lt William Sabol from Phillips Lab's Propulsion Directorate and Maj Dave Swinney from Headquarters AFMC/ST (fig. 19). The combination of these two designs received high marks because the final emblem took into account three fundamental concepts: the origins of the unified lab, a broad range of missions, and the clear implication of the new lab's focus on the future.[10]

The simplicity of the new emblem made it very attractive and appealing. On the left side, five stars arranged vertically serve as a prominent reminder of AFRL's origins. Four of these stars honor the heritage of the four former laboratories— Phillips, Rome, Wright, and Armstrong—and the fifth represents AFOSR. The bright, guiding star in the upper right signifies the lab's constant striving to achieve greatness. At the center of the shield, a three-dimensional triangle traveling to new heights "represents a marriage of aircraft, missile, and spacecraft." The Institute of Heraldry officially described the three-dimensional craft as "silver and white in color to give the impression of quick flight and straightforward in design to highlight the constant and steady advances enabled by our research and development. The craft is pointing to the heavens as it rolls back the night. Thus, through research and development, the light of understanding replaces the darkness of ignorance."[11]

Figure 19. AFRL Emblem

Pleased that the new emblem dignified the heritage, spirit, and mission of AFRL, General Paul on 23 April sent a letter requesting approval to General Viccellio, who quickly concurred. Two weeks later, the Air Force Historical Research Agency at Maxwell AFB, Alabama, advised General Paul that the emblem met the requirements of AFI 84-101, *Historical Products, Services, and Requirements,* which governs heritage guidelines and requirements for all Air Force units. After the chief of staff of the Air Force gave his approval, an artist at the Institute of Heraldry at Fort Belvoir, Virginia, completed the design and artwork of the official AFRL emblem in five months. The original drawing and a final letter of approval authenticating the emblem was then sent to the Air Force Historical Research Agency for permanent retention and safekeeping. AFRL received a copy of the original drawing to serve as the template for making unit flags and any other reproductions of the emblem (i.e., unit patches, decals, signs, and other emblem facsimiles). Prior to approval, no one could officially use the emblem because of the possibility that some minor changes might have to be made during the final design process. As it turned out, that never became a problem. The final, official emblem released in the fall of 1997 was an exact replica of the one submitted by AFRL in the spring.[12]

Road Shows

After General Paul received approval from the secretary of the Air Force in November 1996 to proceed with the formation of the single lab, one of his top priorities was to devise an effective and timely way to keep the troops informed about progress toward that goal. Such a dramatic change in the organization would certainly bring with it a reasonable degree of anxiety and trauma to the lab's large workforce. On the one hand, Paul wanted to reassure people that the creation of the new lab did not pose a threat to their jobs, at least in the near term. On the other hand, he did not try to hide the fact that over the next few years as part of phase IIB, personnel cuts would occur in the support areas (but not for line researchers). He was up front about this, realizing that cuts would come as part of DOD's overall downsizing, regardless of whether the new lab stood up or not. However, he also realized that he could give up vacant positions from the four labs' unit-manning documents and not transfer them to the new AFRL document. Retirements, offers of early outs, and individual job changes to other organizations would also help reach the reduction quota by FY 2001.[13]

Paul also strongly believed in keeping workers informed as events unfolded rather than confirming an event after it happened. Consequently, he wanted to put together some sort of matter-of-fact communications system to give everyone an opportunity to ask questions and receive accurate and up-to-date information on all aspects of the lab reorganization. Above all, employees needed to hear these progress reports directly from the commander—the one person who could best convey a high degree of credibility on a wide variety of lab topics. One of the fundamental principles of command entailed communicating up and down the chain in an honest, timely, and effective manner. Furthermore, because sharing information on a regular basis greatly affected unit morale, it was one of the most important jobs of any leader who depended on his line troops to accomplish the mission.[14]

General Paul took a two-pronged approach to sharing why and how events evolved as they did during the implementation of the laboratory. As mentioned earlier, one of his first methods involved the use of modern technology—setting up a web site—to get the word out quickly and effectively. For the most part, the web site proved very useful because (1) the commander himself wrote messages furnishing timely and truthful information to lab employees spread out from one end of the country to the other and (2) every worker who accessed the site could send questions electronically to the commander, who promised to answer via E-mail.[15]

Certainly, the web site was a valuable source of factual information, but it had one unavoidable drawback, typical of the age of computers—impersonality. Rather than abandon his electronic information center, Paul decided to complement it with what he referred to as road shows—personal visits to each lab, where he would tell people face-to-face exactly what was going on with the lab reorganization. Determined to keep the lab-restructuring process open, he placed a message on his web site explaining the purpose of the road-show briefings: "I believe it is important that you hear the words that go with these charts [on lab restructuring] from me. Additionally, I want to hear from you." Indeed, many lab employees had never seen the commander and would not have recognized him in a crowd of three. Geographic separation and a horrendous work schedule that tied him to Wright-Patterson and Washington, D.C., denied Paul sufficient opportunity to interact regularly with the troops at all levels. Even when he did visit lab sites around the country, he spent most of his time with lab commanders, directors, and other high-level officials who always had the most pressing issues to discuss with him. This left little time, if any, to meet with middle management, junior officers, and enlisted troops—a situation he wanted to correct. Paul therefore made a New Year's resolution to use his road shows to visit all lab sites during the first few months of 1997.[16]

The general made good on his promise, and the road shows paid big dividends to him as well as to lab workers. In February he went to Armstrong Laboratory in San Antonio, following that visit with eight other stops, the last of which took him to Phillips Laboratory at the end of March (table 12). Despite the similarity of the road-show presentations, Paul tailored his briefings to address issues pertaining only to particular sites. Before he left Dayton, he instructed members of his transition staff to solicit questions from employees at the upcoming site, which gave them time to research and obtain answers beforehand. Thus, Paul could provide the most complete information possible and avoid the usual military school response of "I'll get back to you on that."[17]

Consisting of four major parts, a typical road-show briefing provided what Paul referred to as "big-picture highlights" of the status of the new lab. He usually led off with several charts that explained the reasons for establishing a single lab by pointing out the combination of influences at work—principally, the congressionally driven Authorization Act of FY 1996 and *Vision 21*, which called for consolidating labs as much as possible to reduce overhead and infrastructure. This background information also addressed defense planning guidance

Table 12

Road-Show Schedule

Date (1997)	Research Site
13 February	Armstrong Laboratory
3 March	Wright Laboratory/Armstrong Laboratory (local)
7 March	AFOSR
13 March	Hanscom (RL/ER and PL/GP)
14 March	Rome Laboratory
21 March	Wright Laboratory/Munitions (Eglin AFB)
21 March	Armstrong Laboratory/EQ (Tyndall AFB)
27–28 March	Phillips Laboratory (Kirtland AFB) and PL/RK (Edwards AFB)

that required a 35 percent manpower reduction over the next few years, all done to reduce costs, streamline the laboratory acquisition system, and eliminate technology seams. Paul pointed out to all employees at each site the inefficiency of this loose distribution of technologies and the need to establish a more centralized organization that could better control and focus all technology efforts.[18]

Secondly, General Paul talked about the phased implementation schedule, emphasizing phase I events (approving the OCR, reassigning the four labs from the product centers to AFRL, etc.) leading to the interim stand-up of the lab in April. He also described that phase's game plan of weighing recommendations from 13 task groups and then using that information to make decisions on the final structural setup of the new lab. That also played an important role in phase IIA (along with creating a new unit-manning document, selecting tech directors, etc.), requiring implementation before the final stand-up in October. Thirdly, Paul focused on the transition process, which identified the transition team and various tech-group leaders, as well as their responsibilities to and interactions with the Corporate Board.[19] Fourthly, General Paul then looked several months ahead, detailing the overall plan for phase IIA (the first set of road-show briefings took place over two months and prior to the start of phase IIA).

At these road shows, Paul always passed on new information as it became available and always set aside sufficient time to answer questions—the latter representing, in many ways, the most satisfying part of the entire briefing. This face-to-face forum gave all employees an opportunity to ask the one person at the top some very pointed questions and allowed them to vent frustrations brought on by the uncertainty of the new lab's future. But after seeing and hearing General Paul explain firsthand all the positive attributes of the new lab and give straightforward answers to their questions, workers at all levels gained a renewed sense of optimism about the system. The general's calm demeanor, as well as his sincere, candid answers to tough questions, relieved a great deal of stress and tension that existed prior to his visit. Although some employees continued to complain and resist change, most of them appreciated the commander's determined effort to bring the story to them.[20]

As an additional benefit of the road shows, Vince Russo, who accompanied General Paul, drew on his expertise as lab-transition director to pass on detailed planning information to the senior leadership at each site. These very productive meetings also allowed the site leadership to tell Paul about specific lab-restructuring issues that affected their directorate. Paul treated these encounters as good listening sessions in which he could obtain a different slant from the field on a wide range of issues and suggestions for possible integration into phase IIA. Most such conversations concerned options pursued by the various task groups, which would play a key role in determining the precise structure of the new lab.[21]

Generally, these get-togethers followed an informal, no-holds-barred format to encourage participants to say exactly what they thought. For the most part, Paul valued the objectivity of the feedback he received from employees at the various research sites and acknowledged that it positively affected his thinking on how the new lab organization should proceed. For example, members of Phillips Lab's Propulsion Directorate at Edwards AFB passed on to him some "very, very strong feelings" about propulsion becoming a separate directorate under the new lab. Paul took this type of information back with him to Wright-Patterson to think about and share with the Technology Task Group and Corporate Board before making final decisions. In this case, after weighing all the pros and cons, he agreed with the propulsion people at Edwards, using their information, along with other data, to help him decide to set up a separate Propulsion Directorate as an integral part of the new laboratory organization.[22]

Most observers credited the success of the road shows to the free exchange of information that took place between the commander and the workforce at each research site. Paul promised that a second round of road shows would start—probably in the summer (although this did not occur until after the October stand-up)—as soon as he and his staff had "hammered out the details of the Phase II AFRL organization." But these briefings represented only one of many important sources of suggestions for putting the new lab together. For example, the 13 internal task groups set up by Russo's transition team as well as the Grassroots Review Board also flooded him with opinions. Additionally, the IAB and yet another review body, the Lab Alumni group (referred to as the Graybeards), furnished Paul with an outsider's view on how the lab should change its ways.[23]

Notes

1. Maj Gen Richard R. Paul to Headquarters AFMC/XP, letter, subject: Organization Change Request Package, 19 January 1997, with attached OCR; and Maj Gen Michael C. Kostelnik, AFMC/director of plans, letter, subject: Single Air Force Laboratory Organization Change Request, 3 February 1997.

2. Paul letter; and briefing, Maj Gen Richard R. Paul, subject: Air Force Single Laboratory, 13 February 1997.

3. Paul letter; and Paul briefing.

4. Special Order GA-9 incorrectly relieved Phillips Lab from the 377th Air Base Wing at Kirtland rather than from its assignment to the Space and Missile Systems Center in Los Angeles. An amended Special Order GA-12, issued on 8 April 1997, incorporated that correction. Paul W. Smith, acting chief, Organization Division, Directorate of Manpower, Organization, and Quality, Headquarters Air Force, to Headquarters AFMC/XPM, letter, subject:

Single Air Force Laboratory Organization Change Request (Headquarters AFMC/XP memo, 3 February 1997), 31 March 1997; idem to AFMC/CC, letter, subject: Organization Actions Affecting Certain Air Force Materiel Command Units, 31 March 1997; and Special Order GA-9, issued by Headquarters Air Force Materiel Command, 1 April 1997.

5. Message, General Paul's web site, subject: Approval and Activation of the Air Force Research Laboratory; on-line, Internet, 10 April 1997, available from http://stbbs.wpafb.af.mil/STBBS/labs/single-lab/updates.htm.

6. Ibid.; and Paul letter.

7. Paul message; Paul letter; and Ms. Wendy Campbell, interviewed by author, 11 June 1998.

8. Paul message; Paul letter; Col Allan R. Nelson, chief, Organization Division, AFMC, to AF/XPM et al., staff summary sheet, subject: Single Laboratory Organization Change Request, 11 February 1997; and Campbell interview.

9. Capt Chuck Helwig, interviewed by author, 3 February 1998; and Maj Gen Richard R. Paul, interviewed by author, 2 March 1998.

10. Maj Gen Richard R. Paul to AL/CC et al., letter, subject: Single Laboratory Heraldry Contest, 10 February 1997; idem, staff summary sheet, subject: Proposed Heraldry for Air Force Research Laboratory, 24 March 1997; message, General Paul's web site, subject: Air Force Research Laboratory Heraldry; on-line, Internet, 1 April 1997, available from http://stbbs.wpafb.af.mil/STBBS/labs/single-lab/updates.htm; and Helwig interview.

11. Paul staff summary sheet; and Paul message, 1 April 1997.

12. Maj Gen Richard R. Paul to AFRL/CC, letter, subject: Request Approval of Organizational Emblem—Air Force Research Laboratory, 23 April 1997; and Julian C. Godwin, Headquarters AFHRA/RSO to AFRL/CC, letter, subject: Acknowledgement of Organizational Heraldry, 8 May 1997.

13. Dr. Vince Russo, interviewed by author, 4 February 1998; Campbell interview; and message, General Paul's web site, subject: Single Laboratory Road Shows; on-line, Internet, 19 March 1997, available from http://stbbs.wpafb.af.mil/STBBS/labs/single-lab/updates.htm.

14. Russo interview; and Paul message, 19 March 1997.

15. Message, General Paul's web site, subject: Getting Under Way; on-line, Internet, 11 March 1997, available from http://stbbs.wpafb.af.mil/STBBS/labs/single-lab/updates.htm.

16. Paul interview; Russo interview; and Paul message, 19 March 1997.

17. Maj Gen Richard R. Paul to AFMC/CC et al., letter, subject: Information Briefing on the Single Air Force Laboratory, 10 February 1997; and Paul message, 19 March 1997.

18. Briefing, Maj Gen Richard R. Paul, subject: Air Force Single Laboratory (Road Show I), 13 February 1997.

19. Ibid.; and briefing, Maj Gen Richard R. Paul, subject: Air Force Single Laboratory (Road Show I), 27–28 March 1997.

20. Russo interview; and message, General Paul's web site, subject: Road Show I; on-line, Internet, 17 April 1997, available from http://stbbs.wpafb.af. mil/STBBS/labs/single-lab/updates.htm.

21. Russo interview; and Paul message, 17 April 1997.

22. Russo interview; and Paul message, 17 April 1997.

23. Paul message, 17 April 1997.

Chapter 12

Other Perspectives:
Independent Review Teams

Perhaps Vince Russo explained it best when he asserted that there was "no absolute[ly] perfect answer to the [lab] reorganization." Everyone seemed to have his or her spin on the most efficient way of dismantling the four labs and then building a completely new consolidated laboratory. As part of the rebuilding process, the leaders at the top needed to attract and consider a diversity of opinions on how to proceed. Casting a wide net to catch as many ideas as possible forced decision makers to look and relook at issues that they intentionally or unintentionally neglected to include in their preliminary planning strategy.[1]

Recognizing that different opinions and perspectives strengthened the planning and implementation process, Paul commissioned a number of review teams for the express purpose of critiquing the formulation of the single-lab structure and offering some time-tested ideas on organizational restructuring. The handpicked members of one such group—the Independent Assessment Board—on which he depended for upfront feedback consisted of some very senior and experienced individuals in S&T who had worked, or still worked, in government, industry, and academia. The mixed background of this prominent group's members, none of whom could be dismissed out of hand because of their collective experience and expertise, provided a beneficial "outsiders' " perspective on the new lab. Involved in major organizational restructuring during their careers, they could offer sound, practical lessons learned from their experiences and could identify potential pitfalls to avoid during the forming of the new lab. Gen Robert T. Marsh, commander of Air Force Systems Command from February 1981 to August 1984, chaired the IAB (table 13).[2]

General Marsh convened the first meeting of the IAB at the Hope Hotel at Wright-Patterson on 8 April. Paul, Russo, and Barthelemy also attended to fill in the group on the overall lab-implementation strategy and progress to date. Paul told

Table 13

Independent Assessment Board

	Gen Robert T. Marsh, USAF, Retired (chair)	
Lt Gen Thomas Ferguson	Dr. Gene McCall	Dr. Arden Bennett
Dr. Robert Selden	Dr. Gary Rapmund	Mr. Jim Sennett
Dr. William Welch	Dr. George Abrahamson	Dr. Natalie Crawford
Dr. Anthony Hyder	Dr. Jim Mattice	Mr. Kirt Lewis

the members that he solicited their input as detached and objective observers of the entire lab-restructuring process. He also addressed the overall concept of operation for the lab reorganization and explained the phased approach that would lead to the stand-up of the lab in October. Next, Russo furnished details about the transition office and identified the various task groups and task forces assigned to projects that required resolution before the stand-up. Barthelemy then briefed the board members on the Technology Task Group and the initial recommendations it presented to the Corporate Board concerning the new tech directorates. After the presentations, the members made comments and asked some very specific and difficult questions, some of which Paul, Russo, and Barthelemy could not answer completely. But they took copious notes on all those key points for the appropriate task group to further investigate and find answers.[3]

The tenor of the IAB meeting was very positive. Although some members questioned the benefits of going to one lab, they did not belabor the point, realizing the decision was irreversible. They wanted to help out as much as they could in terms of providing solid practical and political advice on lab restructuring that Paul and his staff could apply to the implementation process. Marsh and the others brought with them a genuine spirit of cooperation, but they also made clear that the greatest contribution they could offer entailed raising tough, uncomfortable questions that other people might avoid.[4]

At this meeting, the board members raised issues and asked questions that generally affected the long-term implications of the laboratory, as opposed to the more immediate steps required to establish the lab. For example, they had strong concerns about how the lab expected to perform world-class research in an environment of declining resources, including people, funding, and facilities. They did not buy into the slogan of doing more with less. If the lab faced hiring restrictions, what kind of plan did it have to inject new blood into the organization so it could survive and thrive? How would the lab consistently, over the next decade, attract the highly qualified scientists it needed to meet the challenges of the Air Force in an era of rapidly changing technology?[5]

Everyone on the IAB agreed that laboratory leaders needed to pay more attention to a well-thought-out investment strategy that supported long-term S&T programs. They had no doubt that AFRL's mission had to satisfy the development of short-term technology that could transition as quickly as possible to sustain and upgrade operating systems already deployed in the field. However, they argued against a myopic vision that stressed only the near-term benefits of advancing technology. By definition, a laboratory had an obligation to invest in high-risk technologies and allow scientists to pursue far-out ideas that could lead to revolutionary systems for winning future wars decisively. Anything less would shortchange war fighters ordered into battle 10 to 20 years hence.[6]

The day after the meeting, General Marsh sent General Paul a letter summarizing the issues and concerns of the assessment team. Questions posed in the letter addressed broad, long-term lab issues: What is the Air Force vision for the new laboratory? What are the rationale and statement for how this organization will positively affect scientific work and provide timely support to its customers? Can the lab achieve "real" savings and not just cost avoidance? How will AFRL become more valuable to the Air Force, contractors, and academia? Marsh requested that Paul and his staff discuss these questions at the next IAB meeting scheduled for 4–5 June.[7]

Marsh's letter also commented on more immediate issues that affected the lab's implementation plan over the next six months. For example, with regard to the projected decline in manpower, the IAB felt that the lab leadership needed to mount "an assault on the hiring freeze and to better articulate the results of the 35% downsizing already taken." Overhead cost and ways of dealing with it to reduce expenses represented another personnel problem that needed resolution early on in the restructuring effort. Marsh's group also advocated shutting down areas "no longer relevant" in both the technical and support areas. As the lab developed its concept of operation, which would rely on a centralized span of control, the IAB cautioned against giving "full autonomy" to the tech directorates. Doing so would leave AFRL open to criticism for creating 10 to 12 minilabs from four big labs, suggesting that the transformation to a single lab was a more superficial than substantive representation of how the Air Force conducted its science and technology.[8]

The second meeting of the IAB revisited issues and concerns brought up at the first meeting. At the opening of the session, General Paul brought everyone up-to-date on the lab's progress since the last meeting and addressed questions raised in General Marsh's letter. For the most part, the IAB agreed to the direction in which AFRL was heading, but each member seemed to have specific advice on how to proceed with the lab implementation. For example, most members thought that AFRL's vision, mission, and goals adequately addressed future Air Force S&T requirements, although one person declared that the vision sounded like "a high school cheer." In addition, some members thought that the vision and mission needed to stress the synergy required among AFRL, the Air Force battle labs, the commercial sector, and universities.[9]

Paul explained that the mission and vision statements were intentionally short and clear but agreed that people inside and outside AFRL needed a better understanding of AFRL's relationship with other organizations; otherwise, the single lab could not complete its mission. The general pointed out that the just-completed AFRL strategic plan—which described AFRL's vision as "the best people providing the best technologies for the world's best Air Force"—contained what the IAB sought in terms of details about mission and vision. But the plan also defined in more practical terms what the global ideas of the vision really meant. The statement used *best people* as an all-inclusive term to refer to the military/civilian government workforce as well as "our internal and external contractors, our colleagues in government, industry, and academia, and our international partners." *Best technologies* meant "those derived through our in-house and contractual programs, those that our commercial and government colleagues develop that are adaptable to Air Force requirements, and those developed by national and international defense and commercial enterprises." Lt Gen Thomas Ferguson, USAF, Retired, a past director of Science and Technology in Air Force Systems Command and an IAB participant, recommended that AFRL needed to stress in its vision that the new lab would do more than develop new technology. In looking to the future, AFRL's vision would have to rely more and more on exploiting and adapting successful commercial technology to accomplish its mission (a concept included in the strategic plan). If the lab took the lead by identifying and investing in 6.1 through 6.3 technology programs, then industry would follow, investing in those technologies that the Air Force wanted. Finally, *best Air Force* referred to assuring that the US Air Force maintained "its position of recognized world-wide preeminence." In this case, AFRL's strategic plan provided the fuller explanation of vision and mission that the IAB wanted.[10]

The IAB also thought that the AFRL leadership had done a good job of structuring the tech directorates so as to take care of the problem of technology seams. Dr. Anthony Hyder of the University of Notre Dame's Aerospace and Mechanical Engineering Department praised the work of the Technology Task Group: "Aligning the directorates was a tough problem and overall an excellent job was done." Another board member commented that "the directorates are well thought out and provide a good basis for the kind of capability based organization that is envisioned." Others were equally complimentary, remarking that "most of the 'seam' problems appear to have been addressed . . . pretty good resolution of structure—not sure how better to do it."[11]

Although IAB's feedback supported the new tech directorate organization as described by Paul, Marsh's group still found plenty of room for improvement. Knowing that, at the time of the meeting, Paul had not made all of his final decisions on the configuration of the tech directorates, the board members sought to influence his decision and strongly advocated setting up information, sensors, and conventional weapons technologies as separate directorates. Indeed, Paul already favored the notion of three separate directorates, and Sensors, Munitions (conventional weapons), and Information did in fact comprise three of the 10 tech directorates when AFRL stood up in October.[12]

One other IAB issue that drew a great deal of attention and provoked much debate involved clarification of how AFRL intended to resolve the personnel-support problem. Dr. Hyder agreed with General Ferguson's statement that "support staff should not be left at present manning levels," pointing out that no clear plan existed to reduce overhead. This left the impression that overhead positions still characterized the status quo, which tended to weaken the argument that fundamental changes would occur through personnel reductions with the establishment of the new lab. In addition, the IAB remained unconvinced that decentralizing support work at the directorate level, as opposed to centralizing support operations at headquarters, was the right way to go. Under a decentralized plan, AFRL ran the risk of increasing its support staff—a position it did not want to have to defend after the lab stood up.[13]

Hyder's contention that "there was no plan in place to reduce overhead" was not completely on the mark. AFRL leaders did have a plan—but not for immediate execution. For example, no one doubted that the ongoing 35 percent overhead reduction would eventually drop the lab's authorized personnel number below six thousand prior to the deadline of FY 2001. However, a large share of those cuts would begin as part of phase IIB after the stand-up of AFRL in October. Also, although manpower savings comprised only a small part of the total personnel-reduction figures, those savings would occur when the four labs' command sections disappeared in October. Other personnel savings would occur afterwards (e.g., 509 non-R&D Armstrong Lab positions would come off the books as part of phase IIB). Furthermore, under AFRL General Paul had empowered each tech director to reduce overhead/support staff by tailoring the size of the support staff for the particular directorate, depending on local needs and requirements. Finally, by centralizing the XP directorate at headquarters, the tech directorates' XP support staffs would shrink over time as former XP workers moved back to fill S&T positions. How all this would play out in terms of the exact number of support staff assigned to AFRL would not become clear until the end of phase IIB in FY 2001.

In some ways, IAB functioned as devil's advocate by challenging the lab leadership to rethink potentially controversial options before making any final decisions. However, the IAB was not an enemy but a staunch ally of AFRL, very much concerned with the future of the new lab and willing to help in any way possible to endorse the lab's mission and core competencies. The board members expressed their willingness to come back at any time to help the AFRL commander sell the new lab internally and externally after they completed their business at the end of June. They also promoted the value of AFRL by contacting senior Air Force and DOD leaders, contractors, the Air Force Association, and members of Congress to stress the importance of S&T to the nation.[14]

Furthermore, the IAB provided an independent assessment of future AFRL briefings going up the chain of command and worked with and advised AFRL internal staff members on a case-by-case basis to resolve special problems. In other words, the team members would always make themselves available to serve as a sounding board for any laboratory problems that might arise in the future. Finally, the IAB unanimously concurred that lab leaders had to aggressively defend the logic of moving to a single lab and that AFRL leaders, over time, had to demonstrate that utilizing all the advantages of a corporate organizational structure would result in conducting S&T more effectively and efficiently. That had to be a top priority. As one IAB member warned, "Remember that this may be the last stand before consolidation across service lines are [sic] openly discussed." That is, the lab had to have a solid foundation to avoid becoming a target of opportunity for intraservice proponents who wanted to consolidate AFRL with other service labs into a single DOD laboratory.[15]

Thus, the IAB provided a valuable service to General Paul and his staff by providing a detached perspective on a wide spectrum of lab-transition issues. Although Marsh and his colleagues did succeed in causing a great deal of soul searching and in forcing lab leaders to think twice before settling on a policy for completing phase IIA, they did not have a dramatic effect on the outcome of the final decisions. Indeed, Paul, his staff, and the various task groups had already identified and were looking into every issue the IAB raised. Primarily, the board members contributed by focusing on big-picture issues affecting the formation of the single lab and its long-term future rather than detailing how things should be done or what implementation procedures should be followed at the workers' level. Nevertheless, they did note that the success of AFRL depended to a great degree on the basic management principle that "the devil is in the details." The daunting responsibility of working out all the specifics fell squarely on General Paul, his senior staff, tech directors, and over five hundred individuals throughout the organization who diligently worked all the issues connected with the phase IIA implementation.[16]

While the IAB furnished an outsider's view of the laboratory-transition process, the Grassroots Review Board provided an insider's perspective on the planning of the new lab. Vince Russo suggested to General Paul that it would be a good idea to set up an in-house team to acquire midlevel workers' input on how the lab should be structured. Paul agreed, expecting that he would "get some of our most 'unvarnished' advice from this group." He instructed his four lab commanders and AFOSR director to appoint individuals who represented a "diagonal cross section" (from division level down) of the technical and support workforce (military and civilian) to serve on the team. On 11 March Paul announced the names of the 22 individuals selected to the Grassroots Review Board (table 14).[17]

At the Grassroots Board meeting on 1–2 April, General Paul and his staff began by briefing the members on progress made on the implementation plan for the new lab and then invited everyone to ask questions. Other briefings on the first day covered key portions of the reorganization strategy, including technology directorates, plans and programs, support, personnel, and product executives. On the second day, briefings focused on financial management; procurement; integration and operations; corporate information; presence in Washington, D.C.; and headquarters location. The board members then

Table 14

Grassroots Review Board

Armstrong Lab	Phillips Lab	Wright Lab
Maj Warren Zelenski	Dr. Robert Morris	Ms. Pat Petty
Maj Kevin Grayson	Maj Glenn James	Maj T. C. Carter
Ms. Patty Boll	Capt Marsha Wierschke	Lt Col Pat Nutz
Ms. Cheryl Batchelor	Mr. Kevin Slimak	Mr. Charles Stevens
Mr. Jim Hurley	Lt Col Jim Rooney	Dr. Bryan Milligan
		Ms. Mary Kinsella
		Mr. Steven Karacci
Rome Lab	AFOSR	
Dr. Heather Dussault	Dr. Genevieve Haddad	
Mr. Gene Blackburn		
Capt Dale Reckley		
Ms. Rosanne Loparco		

gave their initial impressions of the lab-implementation plan to General Paul and his staff.[18]

Most board members noted that many of the briefings focused on broad ideals but often lacked the necessary details to show how these visions would apply to the workforce. For example, briefers used *customer* and *support* repeatedly but without a clear understanding or exact definition of these terms. Who, specifically, were AFRL customers? Who were the support people, and what was their function in the new organization? The board considered such issues important for accomplishing the overall downsizing plan that would affect individuals' jobs and the ability of AFRL to meet its mission.[19]

Downsizing drew a great deal of attention and prompted a number of questions. For example, the personnel-reduction plan supported the notion that normal attrition (retirements) would take care of the projected cuts in personnel over the next few years. But nothing spoke to the specifics of how that would work—in terms of numbers of people who would leave. Because all personnel reductions would come out of the support areas, board members did not believe that retirements could meet personnel-reduction goals by FY 2001. At that time (April), no one had hard numbers representing real people who would retire from the support areas. Thus, the Grassroots group inquired how anyone knew that attrition would meet the goals of future personnel cuts. Since no personnel reductions would come from the core technology areas, the team asserted that no one could realistically expect that a sufficient number of people in the support areas would retire to meet the 35 percent reduction quota required by FY 2001. General Paul accepted this observation and acknowledged that he and his staff needed to take a much closer look at this issue.[20]

Another important aspect of downsizing, designed to reduce support and strengthen core technology positions, focused on the structuring of the new Plans and Programs Directorate. The overall strategy called for fewer individuals to work in XP positions at the 10 tech directorates since XP would become a more centralized function, located at AFRL headquarters. Most XP positions (considered support) required a degree in science or engineering. As part of the reduction process, many individuals in XP would move out of XP for "reintegration" into S&T jobs assigned at the tech-directorate level. Under that system, support positions would decline while core technology positions would remain unchanged or increase—an unrealistic approach, according to Grassroots members. They doubted that a midlevel or senior person—especially in a supervisory position—who had worked in XP for the past five, 10, or 15 years would care to move down to work S&T at the directorate level. Not only might such a move interfere with a person's career-progression plans in XP, but also it would call into question the effectiveness of a person who had not worked in his or her scientific or technical specialty in a number of years. Although General Paul did not have a good answer for the board members, he once again welcomed their feedback, indicating he would carefully reevaluate his options on this issue.[21]

One other concern of the Grassroots Review Board had to do with the support structure at different geographic locations. At the time of the April meeting, the setup of the support offices remained unresolved, some people favoring a system centralized at AFRL headquarters and others preferring a decentralized arrangement at each of the tech directorates. In either case, the members had concerns about the apparent lack of effective planning provisions to take into account support functions at locations where a directorate did not exist. For instance, no plan was in place for support functions to reside at Edwards AFB to assist rocket-propulsion work. Board members argued that an isolated location, removed from its directorate and AFRL headquarters, needed sufficient support personnel working on site and providing day-to-day services to keep the operation running smoothly. Otherwise, "shadow" organizations would emerge, whereby scientists and engineers would end up performing all types of support work instead of concentrating on their S&T mission. The board believed it fallacious to think that cuts in support personnel would lead to corresponding reductions in the support workload, predicting that the same amount of staff work would exist and that a large share of it would move to the tech directorates. Thus, the remaining support people would have a bigger workload, and some of the support overload would become the responsibility of scientists and engineers.[22]

Paul declared the first meeting of the Grassroots Review Board an "unqualified success" because it brought a fresh perspective to the planning process. He genuinely appreciated the team members' intensity and independence of thought and their willingness to work until midnight to prepare the outbriefing for the second day's meeting. He thanked them for their "participation, hard work, candor, and dedication" and told them how pleased he was with "the results [they] obtained and presented . . . since they made [him] aware of several issues which had not been considered to date—and which [he and his staff] now [would] factor into [their] deliberations." He informed them that he looked forward to their second and final meeting, scheduled for June.[23]

During the second meeting on 23–24 June, General Paul and his staff updated the team on the same issues presented at the first meeting, and the members then prepared their responses. Jim Hurley outbriefed Paul and his staff on the second day, providing them with recommendations and concerns that amounted to a fine-tuning of issues discussed at the first meeting.[24]

Because the proposed structure and function of XP remained a significant concern not clearly understood by all of the Grassroots members, they recommended adding hard details to the implementation plan that would identify the numbers of people expected to work in each new tech directorate's streamlined XP area. Tech directors needed to involve themselves in shaping the functions and authority of the XP organization at headquarters. Without their input and buy-in, they would see XP as a competitor, which in the long run would weaken the organizational efficiency of the new lab. Furthermore, board members did not have a clear understanding of how many and what procedures would eliminate XP positions at the four labs, of whether the new tech directorates would have any XP positions assigned to them, or of what the projected size of AFRL headquarters XP might be.[25]

Team members repeated their doubts that individuals removed from their XP jobs would automatically move down to accept division-level science or engineering positions. They also voiced concerns about the control and tasking authority of XP at headquarters, feeling that under the old four-lab system the XP shop at AFMC/ST issued far too many taskings with short suspenses and far too many repeat taskings. These taskings, which asked for the same information but in a different format, continually frustrated the XP offices at the four labs. Many of the board members feared that the larger and more controlling XP component at headquarters would generate an even greater number of taskings—all of this happening while the reorganization plan called for assigning fewer people to the tech directorates working XP taskings. This translated to a heavier workload for these employees as well as scientists and engineers, who would increasingly be called on to gather desired information at the tech directorates—certainly not an efficient operation for them. The Grassroots team recommended that the XP shop at headquarters build a corporate-knowledge database to which directorates could regularly add information. Each time XP headquarters needed information, it could first check its own database before requesting it from the tech directorates and thus cut down on the frequency of routine and repeat suspenses to the tech directorates.[26]

The Grassroots team also sought clarification on whom the people working XP taskings at the directorate worked for—the local tech director or central XP at AFRL headquarters? General Paul responded to that question on the spot, stating that all the local-directorate XP components would be a part of and subordinate to the main XP at headquarters. He viewed XP as an organic component of his headquarters, controlled directly by him. Thus, he could formulate and issue one lab-consolidated policy on plans and programs. If XP were decentralized at the directorate level, however, then Paul would have to deal with 10 labs, each going in its own XP direction—something he did not want to happen under the single-lab concept of operations. In short, one voice would direct XP policy.[27]

The Grassroots members also suggested that several personnel issues might damage the morale of the new laboratory. If the implementation plan called for gaining efficiency in AFRL by means of personnel reductions, why did the plan not provide for an equitable method for reductions at all civilian grade levels? Clearly, all the cuts would come in the support area, but within that group, the Grassroots team had the impression that all SES and high-grade positions would receive more protection than would the midlevel and lower grades. By definition, support covered all positions not directly working bench science and engineering, including the commander, his staff, XP, finance, personnel, contracting, and so forth. In his response, Paul pointed out that cuts would occur at all levels and support functions over time, including those in the SES ranks. For example, when three associate directors in SES slots retired over the course of the next few years, Paul considered their positions cuts because he did not intend to replace any of them.[28]

Restriction of career opportunities for civilians posed still another personnel problem that might adversely affect morale. The Grassroots team reasoned that as more personnel reductions occurred, creating a smaller workforce, civilians would have fewer opportunities for career progression. If civilians had to face the reality of limited promotions because of a shrinking workforce, how did AFRL leaders expect to recruit and retain quality workers (or expect younger ones starting out in the job market to invest in a career with AFRL)? To offer more opportunities for career progression, the board suggested making Geophysics a stand-alone directorate under AFRL. For a number or reasons, however, that did not happen.[29]

Board members also advised against doing away with deputy branch chiefs as a way of distributing the workload more evenly. If a branch averaged 25 to 50 workers, the chief would need a deputy to help attend to all the administrative duties assigned to that branch, freeing the chief to focus on S&T requirements. Dr. Daniel vigorously opposed this proposal, believing it violated the goal of the "objective directorate" plan that called for organizational consistency to minimize the size and staff support of each of the organizational components (including branches) of the new lab. Ultimately, with the establishment of AFRL, the 351 branches in the old lab structure became 192 new branches. Retaining deputy branch chiefs would mean a loss of 192 positions dedicated to performing core technology work, the heart of the AFRL mission. When the lab stood up in October, it did not include deputy branch chiefs.[30]

Many of the basic issues raised by the Grassroots Review Board were the same ones that the IAB and General Paul's staff had already identified as important topics needing examination. The Grassroots members, however, proved especially forceful in pushing lab leaders for more precise details and refinement of policy procedures and processes that lacked clarity. General Paul reevaluated all this information, either discarding it or applying it in one way or another to his final decision on all the lab-implementation issues.

Paul never called the Grassroots Board back into session to tell its members how he acted on their comments and recommendations, and they did not expect him to do so. As the commander, he could consult with any person or group prior to or after making his final decision. Considering his heavy workload—making judgments affecting 13 task groups, six focus groups, and three independent review teams, as well as a host of other activities—Paul typically analyzed the data and quickly made prudent decisions to expedite the transition process. Because of time constraints, feedback from the commander often became a secondary issue in the larger scheme of things. As one Grassroots participant put it, the most important message sent by General Paul was that "he included people at the working level to provide their input on the laboratory restructuring through the Grassroots team. That showed he cared about the workforce."[31]

Besides the IAB and the Grassroots Review Board, a third review group, known as the Lab Alumni or Graybeards, provided feedback to General Paul on his single-lab implementation plan. Dr. Allan Schell, a former chief scientist at Air Force Systems Command, chaired this group, which included 20 people who had held important positions in the Air Force scientific and technical community. Brig Gen Philippe O. Bouchard, USAF, Retired—former vice commander of Aeronautical Systems Division and commander of the Aero Propulsion Lab, Materials Lab, and Rome Air Development Center in the 1970s and 1980s—was a typical Graybeard. General Paul saw Schell, Bouchard, and other team members as extremely valuable resources whose experience and expertise in Air Force S&T he should draw upon.[32]

The Lab Alumni team met just once, from 9 to 10 June at Wright-Patterson. Of the three independent review groups, the Graybeards had the most criticism of the single lab, challenging that fundamental concept and even suggesting that AFRL leaders seriously consider pausing between phases I and II to reexamine the big issues. But Paul informed them that doing so was not a viable option: "We simply cannot put things on hold without significant repercussions." He and Russo reminded the group that, for all practical purposes, the topic was no longer open for discussion. Rather than delaying the reorganization process, Paul and his staff needed solid suggestions that they could use during the implementation phase to lead to a more efficient day-to-day operation of AFRL. The team responded with several specific recommendations.[33]

The Graybeards suggested that if the new lab intended to compete on the same playing field with the product centers, then its name should give it equal status—hence, Air Force Technology Center. Of course, General Paul rejected that suggestion because the OCR had already approved *Air Force Research Laboratory* for the interim lab. The team also believed that AFRL leaders should rethink the names of the Air Vehicles and Space Vehicles directorates, considering them limiting and misleading because both directorates engaged in technology that fell outside the "vehicles" part of the two directorates' names.[34]

Team members also warned against making the headquarters staff too large. Although Paul's policy called for establishing only a modest staff, the Graybeards feared it could get out of hand and grow quickly. A large headquarters staff had a tendency to become too bureaucratic—a divisive rather than unifying factor. Further, it would nurture an "us" (tech directors) against "them" (headquarters staff) mentality counterproductive to the effectiveness of the lab. Thus, the Lab Alumni strongly recommended populating the Corporate Board with the tech directors, as well as with General Paul and his key staff, and intimately involving the directors in XP planning activities. Abiding by these rules would show unity of purpose by giving tech directors an opportunity to influence the decision-making process on all major policy issues. This would foster a team approach to solving problems rather than relying on a strictly dictatorial way of doing business. In addition, team members viewed the proposal to place product executives in XP to integrate technologies across multiple tech directorates as another example of bureaucratic layering. As for the Graybeards' suggestions to include tech directors as Corporate Board members and not use product executives, General Paul incorporated both of them into the organizational structure of the lab when it stood up in October.[35]

As a way to infuse new ideas and sustain a healthy leadership atmosphere in AFRL, Dr. Schell stressed the importance of rotating personnel between XP and the technology directors. Cross training would build individual depth of experience and expertise in the development of strong and effective future leaders who would best protect the interests of the laboratory. Moreover, career progression would protect the lab by keeping up a pool of quality employees. Hence, laboratory leaders had an ongoing responsibility to lay out "routes by which high performers are rewarded with higher pay and increased responsibility." Finally, the Lab Alumni team believed it essential to establish an AFRL liaison office in Washington, D.C., not only to promote AFRL technical competencies, but also to stem future personnel and budget cuts.[36]

General Paul appreciated the Graybeards' candid feedback and thoughtful recommendations, writing all the team members to thank them for their contributions. He assured them that he and his staff had reevaluated all the key issues they had raised and informed them that he had taken steps to build a stronger and more cooperative process between XP and all the tech directors.[37]

Notes

1. Dr. Vince Russo, interviewed by author, 4 February 1998; and Col Dennis Markisello, interviewed by author, 6 February 1997.

2. Russo interview; Markisello interview; and message, General Paul's web site, subject: Air Force Research Laboratory Independent Assessment Board Review; on-line, Internet, 14 April 1997, available from http://stbbs.wpafb.af.mil/STBBS/labs/single-lab/updates.htm.

3. Russo interview; Markisello interview; Paul message; and Dr. Robert Barthelemy, interviewed by author, 6 February 1998 and 28 January 2000.

4. Russo interview; and Capt Chuck Helwig, interviewed by author, 3 February 1998.

5. Russo interview; and Gen Robert T. Marsh, USAF, Retired, to Maj Gen Richard R. Paul, letter, subject: Concerns/Issues/Comments, 9 April 1997.

6. Russo interview; Marsh letter; and Paul message. For a discussion of the value of long-range R&D, see J. Douglas Beason, *DOD Science and Technology: Strategy for the Post–Cold War Era* (Washington, D.C.: National Defense University Press, 1997).

7. Marsh letter.

8. Ibid.

9. Gen Robert T. Marsh to Maj Gen Richard R. Paul, letter, subject: Independent Assessment Board Review, 24 June 1997.

10. Ibid.; [Dr. Vince Russo,] "Air Force Research Laboratory Strategic Plan," AFRL, May 1997.

11. Marsh letter, 24 June 1997.

12. Ibid.

13. Ibid.

14. Ibid.; and message, General Paul's web site, subject: Status Report; on-line, Internet, 12 June 1997, available from http://stbbs.wpafb.af.mil/STBBS/labs/single-lab/updates.htm.

15. Marsh letter, 24 June 1997; and Paul messag, 12 June 1997.

16. Russo interview; and Paul message, 12 June 1997.

17. Russo interview; Markisello interview; message, General Paul's web site, subject: Single Laboratory Transition Organization; on-line, Internet, 18 March 1997, available from http://stbbs.wpafb.af.mil/STBBS/labs/single-

lab/updates.htm; and Maj Gen Richard R. Paul to AL/CC et al., letter, subject: Grassroots Review Board for the Single Laboratory Effort, 11 March 1997.

18. Dr. Robert Morris, interviewed by author, 16 March 1997; and Maj Gen Richard R. Paul, briefing notebook, "Air Force Single Laboratory: Grassroots Review Board," 1–2 April 1997.

19. Maj Marsha Wierschke, interviewed by author, 15 and 16 March 2000; briefing slides from Major Wierschke's Grassroots notebook, 1–2 April 1997; and message, General Paul's web site, subject: Grassroots Review Panel; on-line, Internet, 14 April 1997, available from http://stbbs.wpafb.af.mil/STBBS/labs/single-lab/updates.htm.

20. Wierschke interviews; and Morris interview.

21. Wierschke interviews; and Morris interview.

22. Wierschke interviews; Dr. Bill Borger, interviewed by author, 27 July 1998; Maj Gen Richard R. Paul, interviewed by author, 2 March 1998; and Wierschke briefing slides.

23. Morris interview; and Paul message, 14 April 1997.

24. Morris interview; and Mr. Jim Hurley, interviewed by author, 17 March 2000.

25. Paul interview; Morris interview; Hurley interview; Wierschke interview, 16 March 2000; and Wierschke briefing slides.

26. Morris interview; Hurley interview; Wierschke interview, 16 March 2000; and Wierschke briefing slides.

27. Russo interview; and Wierschke interviews.

28. Hurley interview; Paul interview; and Col Mike Heil, interviewed by author, 3 March 1998.

29. Wierschke interviews; Morris interview; and Dr. Vince Russo, interviewed by author, 3 February 1997.

30. Wierschke interviews; Morris interview; Russo interview, 3 February 1997; and briefing, Maj Gen Richard R. Paul, subject: Air Force Single Laboratory (Road Show I), 13 February 1997.

31. Hurley interview; and Wierschke interview, 16 March 2000.

32. Russo interview, 3 February 1998; and message, General Paul's web site, subject: AFRL Progress Report—Directorates and More; on-line, Internet, 14 May 1997, available from http://stbbs.wpafb.af.mil/STBBS/labs/single-lab/updates.htm.

33. Russo interview, 4 February 1998; Brig Gen Philippe O. Bouchard, USAF, Retired, interviewed by author, 20 March 2000; Allan C. Schell to Maj Gen Richard R. Paul, letter, subject: Report of the Laboratory Alumni Review Team, 23 June 1997, with attached Laboratory Alumni Review Team Report; and Maj Gen Richard R. Paul to Brig Gen Phil Bouchard, letter, subject: Lab Alumni, 17 July 1997.

34. Schell letter.

35. Ibid.; Bouchard interview; and Russo interview, 4 February 1998.

36. Schell letter; Bouchard interview; and Russo interview, 4 February 1998.

37. Paul letter, 17 July 1997.

Chapter 13

Headquarters: Two Staff Directorates

Establishing the new Plans and Programs Directorate, designed to help unite 10 separate tech directorates into one corporate laboratory, was critical to setting up AFRL. XP's mission involved establishing and managing all processes for planning, programming, and budgeting AFRL's S&T resources. The directorate's long-range corporate investment strategy and integrated sector planning entailed acquiring, managing, and distributing funding to the 10 directorates in order to support and sustain integrated technology programs and thereby satisfy the technology needs of AFRL's customers. That is, XP had to weave a workable strategy that focused on building a comprehensive S&T program, drawing from multidisciplines, in order to develop the core technical competencies needed to maintain the superiority and cost-effectiveness of US air and space systems. To accomplish that goal, XP had to develop S&T programs responsive to the broad, strategic Air Force goals presented in *New World Vistas, Global Engagement*, and *Joint Vision 2010*.[1]

Plans and Programs Directorate

Strategic goals and money drove investment strategy to ensure that AFRL concentrated on identifying and developing advanced technologies that would eventually transition to the Air Force war fighter. A major part of XP's budgetary responsibility involved determining what to invest in, how much to invest, and where to invest to build the technology base. XP rolled up the projected dollar amount submitted in its Program Objective Memorandum each year to higher headquarters. Justification for what specific technology programs should receive funding in the memorandum resulted from a variety of separate planning and assessment processes and their associated documentation, including the Technology Area Review and Assessment, Defense Technology Area Plan,

Science and Technology Spring Review, Technology Planning Integrated Product Teams, Modernization Planning Process, and more. XP also orchestrated AFRL activities to maintain the lead in technology transfer by managing and coordinating Small Business Innovation Research, Independent Research and Development, and Small Business Technology Transfer Research. Furthermore, the directorate collected and evaluated recommendations provided by the Office of Research and Technology Assessment and the Technology Transfer Office; it also provided oversight for AFRL's international activities.[2]

General Paul appointed Tim Dues to lead the XP task group, instructing him and his group to present all possible options on how best to set up the XP office in the new lab. Like other groups' members, the XP group's came from a number of organizations (table 15).[3] On 24 April, Dues and his team presented three XP options to the Corporate Board. Option one called for a centralized XP operation located at Wright-Patterson that, as part of General Paul's staff, would wield a great deal of power and authority in terms of defining lab-wide policy on all issues related to planning, programming, and budgeting of S&T activities. The advantages of this approach included a strong corporate perspective; a single point of contact, allowing XP to respond quickly to customer requests; and increased potential for developing integrated technology programs across all 10 tech directorates. But locating

Table 15

XP Task Group

Mr. Tim Dues (leader), WL/XP	Mr. Garry Barringer, RL/XP
Dr. Richard Miller, AL/XP	Col William Byrne, PL/XP
Dr. David Dinwiddie, PL/XP	Maj Robert Canfield, AFOSR/XP
Mr. Terry Neighbor, Headquarters AFMC/STX	Lt Col James Feine, Headquarters AFMC/STR
Mr. Bert Cream, AL/HRG	Mr. Ken Feeser, WL/XPZ, Support
Ms. Jan Moore, Headquarters AFMC/STX, Secretariat	Mr. Tom Hummel, Headquarters AFMC/ST-SL, Transition Team POC

XP at headquarters would remove it from the tech directorates, where the base technology work took place. It also meant that, compared to their influence in the four-laboratory system, the tech directors would have less say in determining XP policy.[4]

Option two, a modification of the first, called for keeping a centralized XP at headquarters but establishing smaller XP offices at tech directorate sites (i.e., Kirtland, Wright-Patterson, and Rome), which would allow the central XP to delegate more taskings to the site locations and promote more interactions between the field and headquarters. Option three, the opposite of option one, proposed a decentralized XP that mirrored the former four-lab XP structure, whereby each lab largely developed its own XP policy and direction—something not always conducive to creating a unified XP policy that best served the total organization. In the past, each XP assigned to one of the four labs gave its first allegiance to the lab commander. A decentralized system allowed XP to interact more closely with people who had firsthand experience working the technology programs at the directorate level. It also meant that the headquarters XP would require fewer people—a position favored by the Grassroots and Lab Alumni teams. However, from the perspective of headquarters, the AFRL commander would have more difficulty maintaining a tight span of control over 10 tech directorates, and customers would have 10 XP points of contact rather than one, thus increasing the time required to get a definitive response to their questions.[5]

General Paul made no decision on the XP structure at the April meeting, instructing Dues and his team to reevaluate and refine the three options. He asked them to pay particular attention to XP's responsibilities regarding long-range planning, resource allocation, business-process review, international cooperative efforts, and marketing, with an eye toward determining under which of the three options these XP functions best fit. After rethinking all the XP possibilities, Dues and his team at the next Corporate Board meeting came out strongly in favor of a centralized XP as proposed in option one.[6]

By July, after spending a generous amount of time evaluating all the alternatives for structuring the new XP structure, General Paul made his final decision. His highest priority was to make sure the new lab spoke as one voice regarding all XP issues. He did not like the current system, with XP offices located at five different locations—four labs and the Science and Technology Directorate at Wright-Patterson—under five different bosses. This split arrangement proved extremely time consuming in terms of gathering accurate data for formulating and executing XP policy. The labs wanted to protect their interests and tended to drag their feet in responding to the S&T XP, which had the unenviable task of trying to put together XP guidance, strategic planning, and budgeting that would satisfy four independent laboratories and AFMC. Therefore, Paul chose to go with the centralized XP directorate, which would work directly for him as an organic element assigned to his headquarters and would establish the control necessary to meet his one-lab-commander responsibilities to run an effective XP program.[7]

By eliminating the XP shops at the directorate level, General Paul not only could gain the control he wanted over the XP function but also could reduce the support force and build up the core-technology force, believing that one XP organization at headquarters would require fewer people than had worked at the XP offices at the four labs. He planned for the XP workers at the four labs to do one of two things: (1) apply for the vacant positions that would open at XP headquarters and continue their careers in plans and programs or (2) take a scientist or engineer position at the tech directorate level. In either case, it seemed reasonable to believe that the XP support positions would decrease and that the tech directorates would gain by strengthening their core S&T workforce. However, all this would take time.[8]

Although an essential consideration, the XP manpower issue was not the most important factor influencing Paul's decision. He wanted "one central planning shop living together that could do our strategic planning for us and make us more into a single integrated organization as opposed to the way it was done under the old system." The general believed that the old system represented a stumbling block to effective operations because "it made it very difficult for us to plan that way—planning was fragmented." Under the new XP, the 10 tech directors would play an important role in the planning process, but all final decisions on XP matters would reside with the XP director and commander at headquarters. Although definitely less democratic, a powerful, centralized XP would pay off over the next few years by producing a more united and focused laboratory.[9]

Originally, Generals Viccellio and Paul, backed by the four lab commanders, envisioned four product executive officers playing a vital role in the corporate technology-planning process in the areas of air vehicles, space, C[4]I, and human systems. By looking across all 10 tech directorates to make sure the laboratory exploited the right mix of technologies to develop "specific products," the PEOs would assist the commander in making sound budget and investment-strategy decisions. Paul resolved to fill the PEO jobs with experienced colonels and GS-15s, giving each position the appropriate grade and prestige commensurate with its duties. Doing so would arm each PEO with sufficient status and clout to interact effectively with product-oriented customers such as major commands, product centers, battle labs, DOD agencies, and other government organizations such as the National Aeronautics and Space Administration and the Department of Energy.[10]

In their role as the lab's principal strategists and advocates for product-related technologies, the PEOs would constantly find themselves engaged in a delicate balancing act, assessing and developing programs that crossed multiple technology directorates. Working with the commander, the XP director, and tech directors, the PEOs would prove their value by shaping a corporate investment strategy destined to succeed only with the combined cooperation and expertise of many groups. Ultimately, different groups/customers working cross-technologies together in a coordinated fashion would produce the timely generation of software and hardware products for the Air Force's operational commands, giving them the winning edge during war. In particular, advanced technology would make a decisive difference in the Air Force's six core competencies: air and space superiority, global attack, rapid global mobility, precision engagement, information superiority, and agile combat support.[11]

To make all this happen, each PEO had to work closely with the tech directors and XP director to coordinate the programmatics of multiple technical disciplines. For example, the

Space PEO would consult with the Space Vehicles director and all other directors (of Sensors, Materials, Information, etc.) who had technologies applicable to space systems or products. That process would help the PEOs and tech directors sort out the most appropriate technologies to support Air Force Space Command, Space and Missile Systems Center, Space Battlelab, and other customers as needed. As part of this collaborative effort, Paul envisioned colocating as many as 30 AFRL individuals on a full-time basis with the major commands, battle labs, product/logistics centers, Ballistic Missile Defense Organization, National Reconnaissance Office, and other potential customers. They would serve as liaison officers to best promote what AFRL could do to provide quality technology products to these organizations and assist them in providing customer feedback to AFRL. Although working with many agencies outside AFRL, the PEOs in the chain of command would report directly to the AFRL commander.[12]

As with the XP operation, Paul favored establishing the PEO offices as a centralized function in the headquarters. Under the planned command setup, PEOs would directly "task" the tech directors under certain circumstances—something that did not particularly sit well with the tech directors, many of whom viewed the PEOs as competitors and a potential threat. The IAB and Grassroots Review Board agreed, seeing the PEOs as an extra bureaucratic layer with too much authority, which cut away at the tech directors' autonomy and increased the size of the headquarters. According to the review boards, PEO duties needed to be more precisely defined to prevent both an overlap of responsibilities and any infringement on the tech directors' turf. Furthermore, they considered the name *product executive* confusing because it gave the impression of an AFRL asset in competition with Air Force product centers—just the opposite message the laboratory wanted to convey.[13]

By the middle of April, the issue became more heated and controversial, with General Paul receiving negative feedback challenging the basic concept of PEOs and the role they would play in the new lab. Admittedly, he too found it difficult to define the specific concept and roles of the PEOs. Occupying a separate office in the command section, should PEOs have their own staffs, and, if so, how big should they be? Should

PEOs have budget authority over tech programs? Since the XP office and planned PEO offices both would deal with developing corporate investment strategy, Paul suspected that conflicts would certainly arise over who did what: "The more I tried to peel back the onion and get into the details of a concept of operation for how they [XP and PEOs] would really work, the more convoluted the situation appeared to be." Understanding the controversy stirred up by the PEO proposal, on 24 April he directed Col Rich Davis, the Wright Lab commander, to put together a product executives "Red Team" with the specific tasking to review the entire PEO concept of operation.[14]

Col Richard W. Davis chaired the Red Team.

Senior leadership filled 13 positions on the Red Team, including all four lab commanders—Colonel Davis, Colonel Heil from Phillips, Dr. Godfrey from Armstrong, and Colonel Bowlds from Rome. Most of the remaining members of the team served as tech directors in the four labs. After much debate and analysis, they decided to oppose the establishment of key leadership positions exclusively for product executives in the new lab, questioning the value of PEOs operating independently as a staff office at headquarters. Team members strongly believed that future AFRL tech directors, working with appropriate XP personnel, were perfectly capable of handling all the planning and strategic-investment duties. This option, which seemed a reasonable compromise, tended to shift the focus from "independently powerful" product executives to less intrusive sector chiefs subordinate to the XP director. On 7 May, Colonel Davis delivered to General Paul and the Corporate Board the Red Team's recommendation of eliminating the product executives and replacing them with sector chiefs assigned to XP.[15]

General Paul supported this for several reasons. For one, he liked the idea of embedding sector-chief functions into XP: "That resulted in a more streamlined organization, avoided a matrix staff situation, avoided roles and missions confusion between the PEOs and XP staff, and generally seemed to me to be much more workable from a practicable standpoint. . . . As we ran through various scenarios as to how PEOs versus XP staff would work in response to what-if drills for budget cuts, investment strategy formulation, customer interface, etc., I was reinforced that the XP sector chief approach seemed much more viable. So, that's the way we went."[16]

Another critical issue considered whether the XP sector chiefs should "own resources"—that is, directly control a portion of the S&T budget for reallocation to the tech directors for multidisciplinary, cross-directorate programs. Although some people argued that the XP sector chiefs would have limited power and influence if they didn't have direct budget control, Paul disagreed. He believed that all execution responsibility should reside in the tech directorates and wanted to hold his SES and colonel tech directors totally accountable for program execution; therefore, they needed to own the budget associated with program execution. On the other hand, the general believed that the XP sector chiefs' influence would derive from their direct, corporate-level advice to him on S&T investment planning. Based on his past experience as a staff member in the headquarters of two different major commands, Paul felt that a trusted and competent staff member who worked issues from a corporate perspective could have significant influence on a commander's decision-making process. In some cases, a proficient staff member could have even more influence than line managers who "owned resources"—and Paul thought that a set of highly competent and experienced sector chiefs could have the same kind of influence on the AFRL commander.[17]

With this decision, the die was cast. Sector chiefs would neither own nor control budgets; nor would they serve as program managers for cross-directorate programs. Rather, they would facilitate program planning with the applicable tech directors and advise the AFRL commander regarding specific programs to support from a corporate perspective. After the commander had approved a set of multidisciplinary programs, then the program managers and the budget they needed to execute those programs would reside in the tech directorates. Paul would hold his program managers completely accountable for program execution. For multidisciplinary technology-development programs involving more than one tech directorate, the general planned to appoint a lead tech director who would be accountable for execution and who would work with his tech director peers to ensure adequate support of the program. To Paul, this overall arrangement seemed straightforward from a roles-and-missions standpoint and reflected an appropriate allocation of functions between line and staff.[18]

General Paul's decision to place sector chiefs in XP would not eliminate the tension and conflicts between the line tech directors and the sector chiefs on the headquarters staff, but it would best serve the corporate needs of the new lab. As TEO for science and technology, Paul sought and relied upon advice on budget and investment-strategy matters from both the line and staff, expecting differences in their inputs. He considered both perspectives valuable to his decision-making process. As regards budget and personnel-reduction issues, he anticipated that his tech directors would naturally provide him with in-depth information pushing their respective technology disciplines, while the sector chiefs most likely would advocate a multidisciplinary perspective: "While the tech directors need to put on corporate hats, it's hard to expect them not to defend or advocate their own organizations, manpower, or budgets. If they didn't do that they wouldn't be doing their jobs." Thus, the commander had to analyze and use conflicting input from his line directorates and headquarters staff to help him arrive at the best decision.[19]

The decision to replace PEOs with sector chiefs drew a mixed reaction. The four lab commanders remained unconvinced that sector chiefs buried in the XP organization would have the equivalent horsepower of product executives. Paul tried to alleviate that worry by reminding everyone that the sector chiefs would be handpicked colonels and GS-15s who would maintain a high profile in the lab. Because they would become key participants in corporate budget issues, Paul added them as regular attendees to all Corporate Board meetings. This sent a strong message to both the tech directors and sector chiefs that the lab commander highly valued his sector chiefs. Although Paul had no desire to increase the size of the already large Corporate Board, he thought he needed to make the sector chiefs an integral part of the corporate AFRL decision-making process. But some of the tech directors did not enthusiastically embrace the concept of sector chiefs influencing the AFRL commander to reduce one tech director's budget in favor of another's.[20]

Whether the situation involved money, personnel, or tech programs, Paul stood by his original position of not allowing any of the AFRL tech directorates to act independently in terms of programmatic decisions. On the contrary, AFRL's success would depend upon their interdependence. In most cases, no one tech directorate would own all the technologies that went into enabling a system or subsystem, most of which included integrated multiple technologies. For that reason, he insisted on a strong, centralized XP operation to develop a comprehensive corporate investment strategy that cut across organizational lines to make AFRL a "fully integrated lab." To ensure that the integrated part of the lab-planning process became a reality, he believed it absolutely essential that sector chiefs devote *all* their time and energy exclusively to working issues and relating to customers from a product or mission perspective, as opposed to a technology-discipline perspective. Accordingly, the new sector chiefs would lead five critical product sectors: aeronautical systems, space and missiles, command and control systems, human systems and logistics, and weapon systems. The first four sectors mirrored the four existing AFMC product centers, while the weapons sector corresponded directly to product-related activities in conventional munitions housed at Eglin AFB, Florida.[21]

Primarily, each sector chief would bring together many technologies at different tech directorates into a single technology program for demonstration as an integrated subsystem or system for transition to the operational Air Force. After the sector chief and applicable tech directorates had put the technology plan together, the sector chief would maintain adequate oversight to assure the availability of resources (people, money, and facilities) to support the program and see that it progressed on schedule and met customers' needs. Further, each sector chief had to lead the product cell in XP to develop a sound investment strategy that would logically build AFRL's annual budget submission in an integrated way. Because of the fierce competition for dollars, AFRL would have to submit a compelling and realistic budget request each year if it expected the Air Force to allocate the funding to deliver the most advanced technical products to the war fighter. In other words, the Air Force expected AFRL to deliver an integrated S&T budget linked to the corporate Air Force vision and strategy—not 10 individual budgets for the 10 technology directorates.[22]

In working an integrated program, the sector chief would act as a facilitator with the tech directors and customers to build a technology road map and program schedule, as well as become an advocate to AFRL/XP and General Paul to acquire money to support that program. The tech directorate would then execute that program—perform the S&T—according to the approved road map and schedule. Even after program formulation, the sector chief would remain an important cog in the process, responsible for monitoring progress as the directorate worked the program. If the program needed more money as the S&T moved forward, then the sector chief would advocate additional funds during AFRL's corporate budget-planning process. But as Tim Dues explained, "If he [the sector chief] sees a director pulling funds out of a cross-directorate program or starts to see the road map unraveling or getting out of phase where you lose synergy, where you lose efficiency, then the sector chief has to call him on that." If the tech director and sector chief cannot work out their differences, then the XP director or, if necessary, the commander would resolve the problem.[23]

Using sector chiefs would likely improve efficiency because, under the old four-lab system, too many people became involved in the integrated-planning process. For example, if technologies from two different labs contributed to a single integrated program, then two commanders became involved, who, in turn, would go to their appropriate tech director and XP representative to help work the problem. Because someone at the Headquarters AFMC/ST shop at Wright-Patterson also would become part of the process, as many as seven people might work the same problem. Under the new AFRL structure, however, the sector chief—the XP representative at headquarters—would not have to interact with any commanders since the new lab structure eliminated those positions. Instead, the chief could coordinate directly with the two tech directors contributing to the integrated program, thereby reducing the number of people working the problem from seven to three.[24]

Sorting out the new XP and sector chiefs proved a time-consuming and complicated undertaking. However, by the end of July, General Paul had made up his mind about how XP and the sector-chief structure would fit into the new lab. Knowing that many people lacked a clear understanding of how the new XP organization would function, he issued an unambiguous message on his web site:

> Our XP organization is the integrating and facilitating element that will make AFRL a truly single laboratory. It will be the organizational element that will work with our technology directors to define and collaboratively fund 6.3 programs that cut across multiple technology directives. Just as the four lab/XPs facilitated integrated planning within their respective labs for a set of product-oriented customers (e.g., SPOs, [major commands], AFMC centers, etc.), our AFRL/XP will house planning sectors that are also product oriented: an aeronautical planning cell, a space & missiles planning cell, a [command and control] planning cell, a human systems planning cell, and a weapons planning cell. The improvement here is that all five planning cells will be geographically collocated and will work for the same person, as opposed to the 4-lab situation where the 4 lab/XPs are geographically separated and work for different people—a huge difference! Moreover, within AFRL Phase II, any given XP planning cell can tap the *full* spectrum of technologies in AFRL—not just those traditionally associated with a given product (e.g., [Wright Laboratory] technologies for aeronautical systems, [Phillips Laboratory] technologies for space systems, etc.). I'm confident that this new, centralized AFRL/XP—working in full harmony with our technology directorates—will help us develop and present a powerful corporate portfolio for our weapon system–oriented customers such as product centers, [major commands], and battle labs. We will roll all of the various XP product-sector planning activities together via a corporate investment strategy activity within XP. That activity will be accountable for developing our S&T [Program Objective Memorandum], working budget cuts with the technology directors, preparing planning and programming documentation, and assuring our corporate processes for technology investment planning and technology transition are robust and fully implemented across all of AFRL. Finally, we will house activities such as [Small Business Innovation Research], [Independent Research and Development], international cooperative R&D, and technology transfer within the AFRL/XP organization. As I have repeatedly said, AFRL/XP is the key to helping us become a truly single

laboratory as opposed to "nine small labs." The leaders of the new technology directorates and of the new AFRL/XP will be working hard on an AFRL/XP concept of operations over the next few weeks. There is no higher priority in my mind than assuring that we establish a cooperative, collaborative working environment among the AFRL/XP organization and our technology directorates. Success here means success for AFRL, pure and simple.[25]

Dr. Brendan B. Godfrey, AFRL's first director of Plans and Programs

After he settled on the new XP organization, General Paul then appointed a director to lead this important functional area, so critical to shaping the corporate nature of the new lab. In July Dr. Brendan Godfrey, director of Armstrong Lab, was in Dayton to attend a memorial service for the wife of an Armstrong Lab employee. As they drove together to the service, Paul asked Godfrey if he would take the XP job. After thinking about it a few days, he informed General Paul of his acceptance, fully realizing that the position would become one of the most visible and challenging in the entire AFRL.[26]

Paul chose Godfrey for several reasons. Firstly, knowing that the Armstrong director was going away in October, Paul wanted to make sure that Godfrey moved to a comparable job in AFRL. Secondly, Paul considered Godfrey, who had a strong and distinguished track record in the private and military sectors, eminently qualified for the position. In the late 1980s, he served as vice president of Mission Research Corporation in Albuquerque, New Mexico, moving on to become chief scientist of the Air Force Weapons Laboratory, technology director of the Advanced Weapons and Survivability Directorate at Phillips Lab, and director of Armstrong Lab in San Antonio in 1994. According to Paul, "He will have an unmatched perspective in executing the duties of AFRL/XP." At the same time he announced the new XP director, Paul selected Tim Dues as the associate director of Plans and Programs. As Godfrey's right-hand man, Dues would devote all his time and energy to working corporate investment strategy and technology transition processes. Dues also would work very closely with the six sector chiefs to define corporate policy.[27]

During August and September, when Godfrey began preparing the groundwork to put together the XP organization for the lab stand-up in October, it became readily apparent that Plans and Programs would be a much larger organization in the headquarters than anticipated. Originally, the single-lab plan called for the headquarters staff to increase from about 93 to 150 or 160. But according to AFRL's OCR, submitted in August, 183 authorized positions (62 officers, four enlisted, and 117 civilians) were to be allocated for the Plans and Programs Directorate—more people than the entire headquarters staff projected in the spring—to perform all of the XP requirements of the headquarters and 10 tech directorates.[28]

Paul pointed out that no one had exact numbers on how big XP at the headquarters would become during the early planning stages. His approach entailed building XP from the ground up and sizing it according to the duties attached to that function. After several months of study and recommendations by the XP task group and others, it became clear that XP would become the largest element of the headquarters. That seemed reasonable to Paul because after the lab became a product-center equivalent, the AFRL commander would have more responsibilities and duties in the XP area than he had under the old ST structure. By the summer, when all XP positions were matched against job descriptions at the XP division and branch levels, the number 183 did not seem excessive.[29]

The devil was indeed in the details. To handle all the duties and responsibilities associated with the XP operation would take two divisions and six branches. The Technology Transfer and Corporate Communications Division supervised two branches: Technology Transfer and Corporate Communications. Four other branches—Requirements, Planning, Programming, and International—belonged to the Corporate Investment Strategy Division. So even before the lab stood up, one could see clear signs of a growing headquarters—something the independent review boards had warned the lab leadership about several months earlier.[30]

It became clear in the late summer that the tech directorates would house no more XP offices. Small groups of people
would remain at each tech directorate, working XP issues, but
the plan called for the majority of the four-lab XP employees to
return to line positions within each of the new tech directorates. The word also got out that headquarters would urge
volunteers who had worked at one of the four-lab XP shops to
apply for the large number of unfilled positions slated for
headquarters XP. The moves did not involve promotions—only
lateral reassignments that kept current grades and salaries in
place. The early response was not encouraging. For the most
part, people in field locations removed from Wright-Patterson
simply did not want to pick up and move, mainly because of
family considerations and career aspirations. Senior lab leadership believed that most of these people, over time, would
end up in various tech directorate organizations, such as the
Technology Assessment Division or the Integration and Operations Division.[31]

Operations and Support Directorate

Although organizing the Plans and Programs Directorate
undoubtedly raised some very tough issues for AFRL, it was not
the only major challenge the lab confronted. Deciding what to do
with the support functions in AFRL generated an equal amount
of discussion and emotion and, as did XP, produced two basic
perspectives on how to handle the problem. The centralized approach advocated running support activities from headquarters,
while the decentralized approach called for the tech directors to
select and manage the specific support functions they needed.
The two camps gave the Corporate Board a full spectrum of
support options from which to choose.[32]

Unlike XP, a self-contained business highly focused on integrated strategic planning and budgeting activities, Support
consisted of a loose network of numerous smaller offices, each
with its own distinctive responsibilities for supporting AFRL.
These functions ran the gamut from safety and security to
weather support, with many others wedged in between. If, by
definition, Support consisted of all activities outside the line

Col John Rogacki led the Integration and Operations Task Group.

S&T work, then it accounted for approximately 40 percent of the headquarters and four-laboratory infrastructure. This included communications, facilities management, logistics, human resources, contracting, comptroller services, computer support services, small business services, history, intelligence, legal services, commander's action group, administrative services (mailroom, travel, printing, publications, etc.), supply, multimedia services (graphics, photographic support, displays, etc.), maintenance, protocol, and more.[33]

Two teams examined support options—one led by Colonel Rogacki, director of Phillips Lab Propulsion at Edwards AFB, and the other led by Colonel Markisello, vice commander of AFRL. Both agreed that support required a great deal of attention and thought in building the new AFRL infrastructure, and both recognized the unique opportunity they had to streamline it and make it more efficient. Furthermore, the final support system—however it turned out—had to take into account the mandatory personnel reductions of AFRL's phase IIB downsizing plan.[34]

Rogacki charged the members of his group (table 16) with providing an integrated and efficient support program that would optimize each tech directorate's ability to deliver relevant technologies to the war fighter. After reviewing the current lab-support environment, they would conduct a number of brainstorming sessions that would lead to possible options and then reevaluate and refine each of the options. As part of that process, the team had to focus on two key areas: (1) control of support functions (Who would own the support people? Who would prioritize and direct their work? Who would manage the support offices and be accountable for what they accomplished or failed to accomplish?) and (2) location of support functions. The group would select the best support options

Table 16

Colonel Rogacki's Integration and Operations Task Group

Lt Col Terry Childress, AL/HSC/OET	Col Robert Herklotz, AFOSR/CD
Mr. John McNamaia, RL/OC	Mr. Bob Rapson, WL/MLL
Maj Jim Sweeney, RL/XPP	Mr. Carl Ousley Jr., OL-AC PL/RKS
Mr. Vince Miller, WL/FIIC	1st Lt Kathy A. Zukor, AL/HRT
Mr. David Ramey, Strategic Leadership Associates	Capt Scott Jones, SLTT
Mrs. Darlene Shifflett, AFFTC/MQQ	

for presentation to the Corporate Board, which would make the final decision on how best to organize support.[35]

To better understand the current support organization, Rogacki's team members met with Dues, Russo, and Markisello to find out how the command section viewed support. They then used this information to build seven options consisting of different support configurations. For example, one of the options—the strategic business-unit model—reflected industry's view of support. Private companies, which looked very hard at profits, recommended restricting the support force to approximately 10 percent of the total workforce. But not everyone agreed with that approach, many expressing skepticism about applying the 10 percent ratio to military organizations, which had a higher calling to fight wars in defense of the nation. Driven by its mission, a fighting unit had to depend on an extensive support system. (Indeed, to support one infantry soldier in the field required 14 support troops.) So 10 percent simply did not seem to fit a military operation, one observer commenting that it was "just a number," not factored into the final support solution.[36]

After the team narrowed down the seven options to three, Rogacki briefed the Corporate Board on 29 April, recommending a decentralized support operation closely tied in with each of the tech directorates. This plan called for maintaining only a skeleton support office at headquarters, consistent with the idea of keeping that staff small, as well as keeping support functions under the control of the tech directors. The directors would have the flexibility to size and tailor the support force to fit their directorates' special needs and requirements and would have to pay for each support-service position carried on AFRL's manning document.[37]

At the same time Rogacki's group was preparing its recommendations, Colonel Markisello's Support Task Group (table 17) was hard at work developing and refining a centralized support operation controlled and managed by headquarters. Markisello told the Corporate Board that a strong central site offered the best chance for maintaining consistency of support services across the laboratory since a director of support at headquarters could distribute support resources and services equitably—and with a minimum number of personnel—to all the tech directorates. Instead of allowing each tech director to determine the size of his or her support structure, which could vary greatly from one director to another, the Operations and Support director at headquarters could limit the number of personnel working support issues at the tech directorates—an important aspect since AFRL's personnel reductions would come only from support positions.[38]

Table 17

Colonel Markisello's Support Task Group

Lt Col Stephen Vining, Headquarters AFMC/STDR	Maj Paul Elmer, Headquarters AFMC/STDR
Bonnie Moutoux, Headquarters AFMC/STA	Rosemary Andrews, AL/SD
Wendell Banks, AL/CFP	Scott Marshall, AL
Col Jim Ledbetter, PL/CV/DS	Mike Brown, PL/RKD
Louis Michaud, PL/DS-H	Dan Bollana, RL/DO
Dick Rapke, RL/CE	Jerry Gullo, RL/LG
Col Ronald Channell, WL/DO	Ed Candler, WL/DO
Walt Maine, WL/MNP	Lt Col James Garcia, AFOSR/NC
Scott Jones, Headquarters AFMC/ST-SL	

After listening to the presentations by Rogacki and Mark-isello and after further discussing the two recommended support options with the Corporate Board, General Paul made his decision, selecting Rogacki's decentralized approach to setting up the support infrastructure. He did so for a number of reasons. Firstly, he wanted to keep his headquarters as lean as possible, but a centralized support operation would increase the size of the headquarters and its budget to pay for a large section of support workers, especially since the XP directorate showed signs of growing. The general was determined to keep a small support directorate at the headquarters that would define overall policy and facilitate work on internal support matters.[39]

Secondly, Paul recognized that the majority of support activities took place at the tech directorate locations. Under the current organization, support offices of various sizes and functions were physically scattered at all AFRL locations. Moreover, no two sites configured their support services in exactly the same way, each site having unique local requirements, partly because of mission demands. Depending on the location of the tech directorate, the base infrastructure might or might not be able to furnish some support functions to the tech directorates. The directors liked the current setup because they could usually obtain direct, timely, and reliable service from their own local support people or the base infrastructure. They had no inclination to disrupt this system by depending on support service from a headquarters located, in many cases, hundreds of miles away. In short, they wanted to control their support functions. Furthermore, some expressed concern that, under a centralized system, directorates next to the flagpole at Wright-Patterson—because of proximity—would receive preferential treatment from headquarters for support services. Additionally, many tech directors harbored suspicions that a centralized support directorate could easily lead to micromanagement and dictatorial practices enforced by headquarters.[40]

Thirdly, under a centralized approach, AFRL headquarters would have to build a central support fund by taxing each tech directorate, whereas a decentralized system would require each tech director to finance his or her own support activities. Paul reasoned that, under the latter scheme, most tech directors would keep their support costs to a minimum in order to have more money to conduct research.[41]

General Paul liked the idea that each tech director would size, own, fund, and control the support offices for his or her directorate. Further, the fact that decentralization provided maximum flexibility, allowing the tailoring of support operations to meet specific needs, made sense to Paul, who did not want to force-fit a support template on every tech directorate. The general was also a realist. The existence of so many different support offices—many with fewer than five workers—spread over the entire AFRL landscape would make any kind of move to headquarters extremely difficult. Additionally, personnel regulations protected civilians from forced geographic moves unless their job description included a move clause. So Paul informed his tech directors of his goal of significantly reducing the size of the current support staffs, urging them to reevaluate the size of their support operations at least once a year—preferably more often—as part of an ongoing process to reach the desired manpower savings. As support people retired or changed jobs, tech directors would not backfill those positions, which, over time, would help meet the downsizing requirements.[42]

After Paul made his decision, headquarters and the tech directors began determining the location of each support office. By the time of the stand-up at the end of October, they had assigned every support office to one of three organizations: headquarters, a technology directorate, or a central site where two or more tech directorates pooled their resources and shared support functions. At headquarters, General Paul made Colonel Markisello director of the Operations and Support Directorate, which included a select group of the larger support activities: communications, safety, security, facilities management, logistics, and weather. The 16 people (five officers, one enlisted, and 10 civilians) assigned to DS would coordinate and issue broad policy throughout AFRL for the support functions mentioned above. In addition, each of the support groups in DS had responsibility for ensuring that its particular functions were carried out within headquarters. However, DS did not contain all of AFRL's support offices. The remaining ones were assigned on a case-by-case basis to various other groups in the command section.[43]

The tech directors had to decide what support offices they wanted to own separately and what functions they wanted to include in a common support organization, known as the central site. All of the directors at Wright-Patterson and Kirtland agreed to share a large portion of their support resources at their respective sites. For example, the five tech directors at Wright-Patterson formed a site-operations council that assigned support functions to the Wright central site, with the tech directors owning and paying for only selected support positions on their books (e.g., each separate computer-support service office, although consolidated under one directorate, would serve all five directorates by working out of the central site). Obviously, each tech director would finance roughly 20 percent of the total central-site costs. To sort out what support offices would reside on which directorate's manning roster, all five tech directors met to conduct something similar to the NFL draft. One director would run the central site and coordinate all support activities with the other four directors, which would free them to concentrate on S&T issues without worrying about spending too much time attending to support requirements. The two tech directors at Kirtland set up a similar central-site operation.[44]

Not all common support functions ended up in a central site, however. At Kirtland, some support offices assigned to Phillips Lab—such as human resources, safety, security, protocol, public affairs, and others—were split. The two Kirtland tech directors agreed that the safety office serving Phillips Lab should become two separate offices when the lab opened in October, half assigned to the Integration and Operations Division to support the Directed Energy Directorate and the other half assigned to support the Space Vehicles Directorate. Similarly, several months after the AFRL stand-up, Dr. Russo saw to it that the Materials Directorate at Wright-Patterson had its own personnel support office. By having his personnel office down the hall, he could take care of all personnel matters in a more effective and timely manner.[45]

While Paul's staff worked on finalizing the details of all command-section staff offices, several very important personnel changes occurred in the middle of the summer. At the busiest time of the reorganization process, only three months before the phase II stand-up, three of the four lab commanders departed for new assignments, prompting many people to question why the top leaders would leave in the midst of such a radical change when the organization needed them most. Hundreds of taskings remained incomplete, and personnel uncertainties pushed the anxiety level at the lab to an all-time high. Many employees, searching for reassurance and a steady hand on the helm, felt that the lab commanders had abandoned ship to advance their own careers.

General Paul, however, saw things differently. Because of the decision to replace product executives (the lab commanders had become product executives during the interim lab stand-up) with sector chiefs, no high-level positions remained for the lab commanders to fill. In many ways, they had become odd men out. Paul realized in August that the new tech directors had the deepest involvement in preparing the lab for its October stand-up. They and their staffs worked unit-manning and infrastructure issues to make sure the tech directorates shaped up as planned. More and more of the burden and workload passed from the commanders to the tech directors, who would assume control once the stand-up occurred. In fact, one could argue that, during the few months prior to the end of phase IIA, the lab commanders had become mere figureheads. Recognizing their dilemma, Paul encouraged them to move on and accept jobs of higher responsibility with other organizations.[46]

General Paul announced that Col Rich Davis, commander of Wright Lab and a brigadier general as of 1 August, had to move on to the Ballistic Missile Defense Organization because no O-7 positions existed in AFRL. Dr. Keith Richey left his job as director of the Flight Dynamics Lab to take over as director of Wright Lab until the end of phase IIA, and Paul appointed Col Ron Channell as the Wright Lab commander for the same period.[47]

At Phillips Laboratory in Albuquerque, Dr. Earl Good became the new director, replacing Col Mike Heil, who moved on to become inspector general at Headquarters AFMC. Col Bill Heckathorn became the new Phillips Lab commander. In upstate New York, Mr. Ray Urtz replaced Col Ted Bowlds, who left Rome Lab to take a new assignment at SAF/AQ in the Pentagon. Col Fred Foster became the Rome Lab commander. Dr. Brendan Godfrey, who remained in his position as director of Armstrong Lab, would move to his new job as head of AFRL/XP in October.[48]

Faced with the prospect of losing their jobs through reorganization, the lab commanders nevertheless put on corporate hats and rolled up their sleeves to help formulate the single laboratory during some very challenging times. Paul had generous praise for them: "I was proud to be associated with these outstanding leaders. They have my deepest respect and gratitude."[49]

Notes

1. Dr. Keith Richey, former chief scientist, Wright Laboratory, interviewed by author, 5 February 1998; briefing, Mr. Tim Dues, WL/XP, subject: AFRL/XP Task Force: Status Report, 22 March 1997; Maj Gen Richard R. Paul, AFRL Corporate Board notebook, meeting materials, tab 14, "Mission Statement: Air Force Research Laboratory Plans and Program Directorate," 25–26 August 1997; and Robert Ely, "Air Force Consolidates Research under New Lab," *KAFB Focus*, 31 October 1997. See also *New World Vistas: Air and Space Power for the 21st Century, Summary Volume* (Washington, D.C.: USAF Scientific Advisory Board, 1995); *Global Engagement: A Vision for the 21st Century Air Force* (Washington, D.C.: United States Air Force, 1997); and *Joint Vision 2010* (Washington, D.C.: Joint Chiefs of Staff, 1995).

2. Richey interview; Dues briefing; Paul notebook; Ely; and Maj Gen Richard R. Paul to AL/CC et al., letter, subject: Laboratory Vision Statement, 21 July 1996 with attached "Vision Statement."

3. Dues briefing.

4. Ibid.

5. Ibid.

6. Mr. Tim Dues, interviewed by author, 2 March 1998.

7. Ibid.; and Maj Gen Richard R. Paul, interviewed by author, 2 March 1998.

8. Dues interview; Paul interview; and Col Dennis Markisello, interviewed by author, 6 February 1998.

9. Paul interview; and Dr. Brendan Godfrey, interviewed by author, 10 June 1998.

10. Maj Gen Richard R. Paul, telephone conversation with author, 2 April 2000; briefing, Maj Gen Richard R. Paul, subject: Air Force Science and Technology: Product Executives, 2 April 1997; briefing, Maj Gen Richard R. Paul, subject: Air Force Single Laboratory, 13 February 1997; and Maj Gen Richard R. Paul to author, E-mail, subject: Update, 2 April 2000.

11. Paul briefings, 13 February and 2 April 1997.

12. Paul briefing, 2 April 1997; and briefing, Maj Gen Richard R. Paul, subject: Air Force Research Laboratory: Heading Check, 29 April 1997.

13. Paul briefings, 2 and 29 April 1997.

14. Maj Gen Richard R. Paul, interviewed by author, 22 November 1999; Capt Chuck Helwig, ST-SL, point paper, "Product Executive Red Team," draft [April 1997]; Paul E-mail; message, General Paul's web site, subject: AFRL Phase II: A Decision Process Flowchart; on-line, Internet, 25 April 1997, available from http://stbbs.wpafb.af.mil/STBBS/labs/single-lab/updates.htm; and Paul briefing, 29 April 1997.

15. Briefing chart, "Product Executives Task Force Team," n.d.; and briefing, Col Rich Davis, subject: Report of PE [Product Executive] and PE Red Team Subgroups, 7 May 1997.

16. Paul E-mail.

17. Maj Gen Richard R. Paul, notes, subject: Lab Reorganization, 20 April 2000.

18. Ibid.

19. Paul E-mail; and Maj Gen Richard R. Paul, telephone conversation with author, 3 April 2000.

20. Paul E-mail; and Paul telephone conversation, 3 April 2000.

21. Paul interview, 22 November 1999; Maj Gen Richard R. Paul, interviewed by author, 25 April 2000; message, General Paul's web site, subject: AFRL Progress Report—Directorates and More; on-line, Internet, 14 May 1997, available from http://stbbs.wpafb.af.mil/STBBS/labs/single-lab/updates.htm; message, General Paul's web site, subject: Key Leaders for the AFRL Phase II Organization; on-line, Internet, 22 August 1997, available from http://stbbs.wpafb.af.mil/STBBS/labs/single-lab/updates.htm; Dr. Donald C. Daniel to all AFRL, letter, subject: Request for Volunteers for the Air Force Research Laboratory Plans and Programs Directorate, 21 October 1997, with attached AFRL/XP mission statements; and Paul notes.

22. Paul interviews, 22 November 1999 and 25 April 2000; Paul messages, 14 May and 22 August 1997; Daniel letter; Paul notes; Dues interview; and Dr. Vince Russo, interviewed by author, 4 February 1998.

23. Dues interview.

24. Ibid.

25. Message, General Paul's web site, subject: AFRL Phase II Status; on-line, Internet, 31 July 1997, available from http://stbbs.wpafb.af.mil/STBBS/labs/single-lab/updates.htm.

26. Godfrey interview.

27. Paul message, 22 August 1997.

28. AFRL Organization Change Request Package, appendices A–L, August 1997.

29. Paul interview, 2 March 1998.

30. Daniel letter.

31. Godfrey interview; Col Mike Pepin, interviewed by author, 3 February 1998; Col John Rogacki, interviewed by author, 9 June 1998; and message, General Paul's web site, subject: Personnel Allocation Process; on-line, Internet,

16 September 1997, available from http://stbbs.wpafb.af.mil/STBBS/labs/single-lab/updates.htm.

32. Rogacki interview; Markisello interview; and Col Dennis Markisello to author, E-mail, subject: Single Lab History, 3 April 1998.

33. Briefing, Col Dennis Markisello, subject: Guiding Principles for Support Functions [July 1997]; and briefing, Maj Gen Richard R. Paul, subject: Air Force Research Laboratory: Status Update, August 1997.

34. Rogacki interview; Markisello interview; and briefing, Col John Rogacki, subject: Integration/Operations Division Team [March/April 1997].

35. Rogacki interview; Markisello interview; and Rogacki briefing.

36. Rogacki interview; and Dr. Robert Barthelemy, interviewed by author, 6 February 1998.

37. Rogacki interview; Paul briefing, 29 April 1997; and Paul message, 25 April 1997.

38. Rogacki interview; AFRL briefing chart, "Support Task Force Team," n.d.; and Paul briefing, 29 April 1997.

39. Paul interview, 2 March 1998; and Paul message, 14 May 1997.

40. Markisello interview; Paul interview, 22 November 1999; Rogacki briefing; and Paul message, 31 July 1997.

41. Paul notes.

42. Paul interview, 2 March 1998; Markisello interview; Markisello briefing; and Paul message, 14 May 1997.

43. Paul interview, 2 March 1998; Markisello interview; Markisello briefing; Paul messages, 14 May and 22 August 1997; Russo interview; and AFRL Organization Change Request Package.

44. Dr. Bill Borger, interviewed by author, 27 July 1998; Mr. Les McFawn, interviewed by author, 30 July 1998; Russo interview; Capt Chuck Helwig, interviewed by author, 3 February 1998; Paul briefing, 29 April 1997; and Paul message, 14 May 1997.

45. Borger interview; and Russo interview.

46. Paul interview, 2 March 1998.

47. Message, General Paul's web site, subject: Transition from AFRL Phase I to Phase II; on-line, Internet, 4 August 1997, available from http://stbbs.wpafb.af.mil/STBBS/labs/single-lab/updates.htm.

48. Ibid.

49. Paul notes.

Chapter 14

The Final Push

In addition to the two functional directorates at headquarters—Plans and Programs and Operations and Support—a number of staff offices made up other important elements of the AFRL command section. One of these, the Washington Office, evolved as a result of the work produced by the D.C. Presence Task Group, chaired by Dr. Godfrey. This small but very experienced group consisted of Dr. Joe Janni and Mr. Matt Jaskiewicz from AFOSR, Col Mike Havey from Phillips Lab, Col Brendel Kreighbaum from Rome Lab, Mr. Bill Woody from Wright Lab, and Lt Col Jim Rader from Armstrong Lab. This group would explore the feasibility of establishing an office to promote and market AFRL's capabilities, programs, and products to customers in the nation's capital. Essentially a liaison operation, the office aimed to put AFRL on an equal footing with other federal labs in terms of competing for resources.[1]

The Bergamo conference of December 1996 had raised the issue of moving the new laboratory's headquarters to Washington, D.C., in order to enhance visibility and gain the support of Washington power brokers for the Air Force's S&T programs. Generals Viccellio and Paul, however, adamantly opposed such a plan, arguing that everyone had more than enough work to complete on a tight schedule without adding the burden of moving headquarters to a new location hundreds of miles away from Dayton. Viccellio firmly intended to keep AFRL—one of the key elements of his command—close to Headquarters AFMC at Wright-Patterson so he could interact with General Paul, daily if necessary. Similarly, Paul intended to keep his headquarters near the largest technical component of his organization (five of nine tech directorates) at Wright-Patterson and close to his AFMC boss.[2]

Nevertheless, Paul continued to receive inquiries from "product center commanders and others" about permanently locating AFRL headquarters in Washington—a good example of taking too much time to solve 10 percent of one's problems. Wanting to put this matter to rest once and for all so that he and his staff could get on with more pressing lab matters, Paul wrote to General Viccellio in early February, asking him to verify his intentions concerning the lab's location. Viccellio responded in the clearest language possible: "The single lab HQ will be at WPAFB—final decision."[3]

Despite their opposition to moving AFRL's headquarters, they did favor coming up with a plan that would give the new lab as much exposure as possible in Washington. That job fell to Godfrey's group, which held its first meeting on 18 February. Taking the position "If you're not there, you don't play" in the Washington arena, the members set out to define the goals and structure for establishing a strong AFRL presence there. Other federal labs, such as Sandia, Los Alamos, Livermore, and Oak Ridge maintained permanent staff offices in Washington to assure that they received their fair share of DOD's economic pie.[4]

Godfrey briefed his group's findings to the Corporate Board in March, April, and May, reporting that the Washington liaison office should serve as AFRL's sole S&T voice in dealings with the Air Staff, OSD, Congress, and "other Washington-area stakeholders." The latter included Washington-based professional technical societies and research institutes, the Army and Navy, government agencies, major aerospace companies, and universities. The director of the office would serve as the AFRL commander's representative and spokesman for AFRL in national forums and committees. A successful office would garner a larger fraction of DOD funding for AFRL programs and pave the way for the lab's participation in joint-service programs that would leverage Air Force funding for AFRL.[5]

During the Corporate Board discussions, General Paul stressed that the office's director would not have the authority to make independent decisions affecting the future of the lab but would work very closely with him and his staff, as well as the tech directors and their technology program managers. Only after proper internal coordination could they develop a coordinated and consistent long-range policy for AFRL, allowing the D.C. director to promote AFRL's position at high-level meetings in Washington. This did not restrict the director from taking the initiative to probe and investigate potential contacts in D.C. that might benefit AFRL, but final commitments to specific technology programs would take place only after General Paul had consulted with his senior staff and technology directors.[6]

After reviewing all the information he received from Godfrey's team as well as the Lab Alumni group's comment that "the Washington D.C. presence of AFRL is critical," Paul approved the establishment of a Washington office. In July he announced on his web site that the new office would take the lead in "promoting AFRL's interests among our Washington area partners and customers." To ensure that the office had "adequate horsepower," he assigned Dr. William O. Berry, an SES and former director of chemistry and life sciences at AFOSR, as its first director. Paul colocated the new office and its staff of four people with the Secretary of the Air Force for Acquisition (SAF/AQR) organization in the Pentagon to avoid having two Air Force S&T voices in Washington—whether in fact or in perception. He and Dr. Helmut Hellwig, head of SAF/AQR at the time, were in complete agreement about this arrangement and further agreed that to ensure tight linkage between the two offices, Dr. Berry should also serve as SAF/AQR deputy to Dr. Hellwig. Convinced that the Washington presence would have a positive effect on AFRL's future, General Paul told the AFRL workforce that he was excited about the "doors to be opened by our D.C. office."[7]

Paul then announced that eight other staff offices—mandated by Air Force organizational policy—would round out the remaining organizational elements of the headquarters command section. The largest of these, the Human Resources Office, with 19 employees, combined the traditional functions of manpower (spaces) and personnel (faces). The two smallest organizations—the Corporate Development Office, responsible for corporate, organizational, and employee development activities, and the Reserve Affairs Office—each had only two civilians assigned. A new function, the Corporate Development Office, focused resources and training on organizational skills (e.g., leadership, teamwork, participatory decision making, etc.) for AFRL's workforce.[8]

The Corporate Information Office (CIO), consisting of one officer and four civilians, would define and acquire AFRL's corporate information infrastructure. General Paul wanted a CIO that would take the lead in linking modern information-technology systems (computer networking, E-mail, database management, standardized software and hardware, etc.) across the lab: "How we manage information and business communications in our new lab will be crucial to our success." Despite CIO's small size, Paul envisioned a decentralized operation in which the tech directorates would house and manage most of the office's staff. Furthermore, CIO was not just a local agency, Congress having passed in 1996 the Information Technology Management Reform Act (the Clinger-Cohen Act) that established the position of chief information officer for all government agencies.[9]

Other offices assigned to the headquarters included the Commander's Action Group, which worked special projects for the commander on a case-by-case basis; the Executive Services Office, which managed all command administrative activities, including correspondence tracking, protocol, awards, promotion and retirement ceremonies, supply purchasing, and so forth; the Comptroller Office, responsible for the financial management of AFRL; and the Contracting Office, manned by Headquarters AFMC personnel matrixed to AFRL. The latter office advised the AFRL commander on acquisition policy; the awarding of contracts, grants, and cooperative agreements; and other acquisition transactions. Since AFRL "outsourced" 75 percent of its budget to private industry and universities, contracting personnel had to assure the legality, efficiency, and responsiveness of all contracting actions.[10]

Finally, General Paul announced the establishment of eight research site detachments at Brooks, Edwards, Eglin, Hanscom, Kirtland, Rome, Tyndall, and Wright-Patterson Air Force Bases, with a colonel designated as site commander for each detachment. Their "Series G orders" gave these colonels, who would report directly to General Paul, authority commensurate with their positions, particularly regarding *Uniform Code of Military Justice* authority to deal with military disciplinary issues promptly and decisively. The site commander staffed the orderly room, which served the entire AFRL military population at the installation, in addition to serving as director, deputy director, or division chief of a tech directorate at the detachment location. This arrangement would save manpower and reflected the fact that the site commander's duties were not full-time jobs.[11]

None of the offices assigned to headquarters would become official until after the stand-up. In order to obtain authorization for that event, General Paul and his staff had to submit an OCR for approval by the AFMC commander and the Air Staff—specifically, the chief of staff and the secretary. That document had to respond to seven basic questions. *(1) What is the proposed action?* The action entailed inactivating four laboratories and consolidating them into one laboratory consisting of 10 technology directorates. *(2) Why is the action needed?* Congressional action embodied in section 277 of the National Defense Authorization Act for FY 1996 required the Air Force to consolidate and restructure its laboratories. *(3) What is the structure of the new organization?* Obtaining all the data to answer this question proved extremely time-consuming and arduous (see below).[12] *(4) How does the structure compare with the standard structure and nomenclature?* No standard structure for AFMC laboratory functions existed because the mission requirements made each current laboratory organization different. *(5) What are the potential impacts on other organizations?* The formation of AFRL would minimize technology seams that existed in the four-lab organization and, in the process, would reduce the single lab's infrastructure operating costs. Further, each of the AFMC product centers, previously serviced by only one of the four labs, would now gain the support of the entire one-laboratory enterprise. *(6) Why is it better?* The new lab would reduce management and support overhead functions, which would eventually decrease the number of authorized positions assigned to AFRL. *(7) What impact does the organization request have on unit history?* Each of the four labs would prepare a close-out history, and historians in the new lab would be responsible for writing AFRL histories.[13]

As mentioned above, question number three took some time to answer. The most tedious undertaking involved assigning specific authorization numbers for officers, enlisted personnel, and civilians to match each of the newly created staff offices and functional directorates in the command section, as well as the technology directorates. This process had started back in the spring with Dr. Russo and his transition team reviewing the four labs' unit-manning documents to begin the process of locating each position in the new organization. By July this effort had intensified greatly, because they had to include that vital information in the OCR (fig. 20). In terms of the organizational change that took place, one can easily see a flattening of the AFRL structure, compared to the old four-lab structure, as well as the elevation of the lab to a "tier one" organization in the AFMC organizational structure, on par with the product, test, logistic, and specialty centers (figs. 21 and 22).[14]

By the end of July, General Paul and his staff had documented all the information required in the phase II OCR and sent it forward to Headquarters AFMC for coordination. General Babbitt approved and signed the OCR on 7 August and sent it to the Air Staff at the Pentagon on 8 August. Paul predicted that it would take 60 to 90 days to move the document through the staffing process, so he expected to hear that AFRL could officially stand up by 1 October.[15]

By the middle of September, Paul had received no word on the OCR except that it was working its way through the Air Staff. At that time, he announced that each of the new tech directors was hard at work implementing the details for configuring the new directorates' divisions and branches. Lab guidance required the directors to build the divisions and branches in accordance with AFRL's mission and then assign individuals who formerly worked for the four labs to the new slots in the tech directorates. This laborious process, which tried the patience of everyone involved in the exercise, was just one of the growing pains that lab people had to adjust to and endure. Even though each person had an assigned slot when the lab stood up, they realized that those assignments might change over the next few months. All of the personnel offices throughout the lab were closely reviewing each person's skill requirements to make sure the right person matched up with the right job.[16]

The first of October came and went without any word on the OCR. On his web site, Paul posted a status message entitled "What's the deal?" explaining to the lab workforce that, although the Air Staff was taking longer than anticipated, he knew that the OCR had reached the vice chief of staff's office.

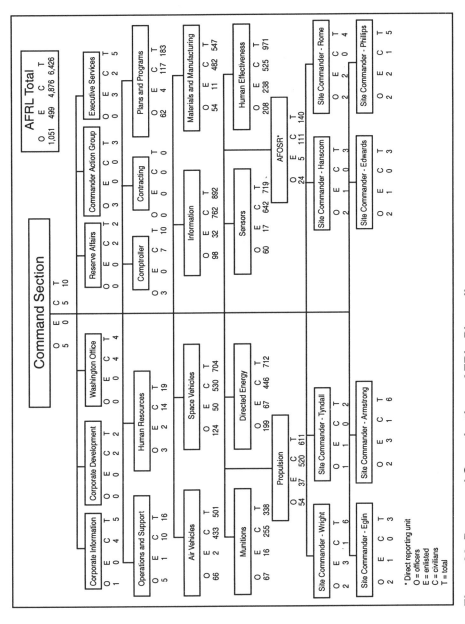

Figure 20. Proposed Organization—AFRL, Phase II

259

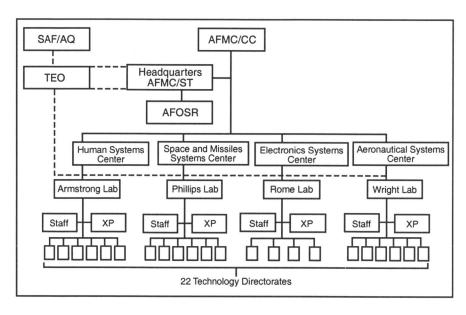

Figure 21. Pre-AFRL S&T Organization

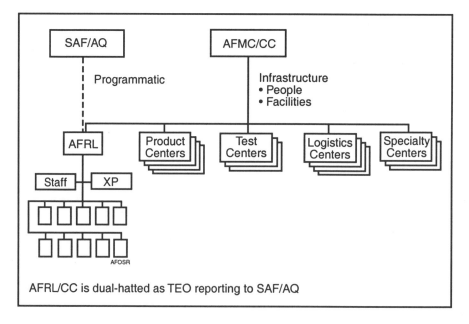

Figure 22. Air Force Research Laboratory

After he signed it, the OCR would then go to the chief's and secretary's offices. Brig Gen Larry W. Northington, who headed the AF/XPM shop in the Pentagon, assured Paul that it simply took time to push the OCR through the paperwork process, especially since the Air Force would soon transition to a new chief of staff. Although Northington saw no problem with the OCR package that could "derail it," Paul nevertheless made a trip to the Pentagon to go over the package with Northington and ensure that he had no lingering questions.[17]

Two weeks later, Paul wrote, "It's official!" Secretary of the Air Force Sheila Widnall, just a few days away from leaving her post, approved the AFRL reorganization on 22 October. The next step called for the Air Staff to send a letter to AFMC/XPM to issue the G-series orders inactivating the four labs and establishing AFRL and its sites, detachments, and operating locations. Allowing some time to complete all the paperwork, Paul and AFMC set the effective stand-up for 31 October, after which all AFRL organizations would start using their new office symbols. After nearly 11 months of difficult work and long hours invested by hundreds of people at all levels of the laboratory workforce, Paul was jubilant, knowing that the new lab was a reality.[18]

Although looking optimistically to the future, General Paul took the time to remind everyone of the lasting contributions and rich heritage of the four original laboratories:

> In the coming weeks, Armstrong, Phillips, Rome and Wright Laboratories may hold ceremonies or other events commemorating their great histories. I will leave the decisions as to whether to hold such an event, and how it should be conducted, to each of our lab sites so they can tailor it to their own desires. As inactivated units, the labs will retain their history, honors, and lineage. In order to further preserve the histories of the laboratories, we have taken the step of naming four of our Research Sites after them. My hope is that our laboratory heritage is not lost in the new AFRL.[19]

Col Jacob Kessel, the same AFMC chief of Manpower and Organization who cut the AFRL interim orders in April, issued Special Orders GA-1 and GA-2 on 29 October 1997. The first order activated the Air Force Research Laboratory and eight research site detachments, effective 31 October 1997 (table 18). Because of the tech directorates' geographical separation from

AFRL headquarters at Wright-Patterson, the order also activated four technology directorate detachments (table 19). The second special order (GA-2) officially inactivated Armstrong, Phillips,
Rome, and Wright laboratories, effective 31 October 1997.[20]

The next day, Colonel Kessel notified General Paul that the
naming of all the command section staff offices, functional
directorates, and tech directorates identified in the OCR had
received approval. Kessel further noted that the AFRL vice
commander would also serve as the AFRL inspector general,

Table 18

AFRL Research-Site Detachments

Detachment	Location
1	Wright-Patterson AFB, Ohio
2	Tyndall AFB, Florida
3	Hanscom AFB, Massachusetts
4	Rome, New York
5	Brooks AFB, Texas
6	Eglin AFB, Florida
7	Edwards AFB, California
8	Kirtland AFB, New Mexico

Table 19

AFRL Technology-Directorate Detachments

Detachment	Location
9 (Space Vehicles)	Kirtland AFB, New Mexico
10 (Information)	Rome, New York
11 (Munitions)	Eglin AFB, Florida
12 (Directed Energy)	Kirtland AFB, New Mexico

that 509 Program 8 personnel authorizations would transfer from AFRL to the Human Systems Center, and that the four product executive positions originally planned would be eliminated.[21]

An official military ceremony to stand up the Air Force Research Laboratory took place on 22 October during an AFMC commander's conference at the Dayton Convention Center. As a small group of senior military and civilians watched, Gen George T. Babbitt, AFMC commander, passed the newly approved AFRL flag to Gen Richard R. Paul, who took command of the Air Force's first unified laboratory. This marked the end of an era of multiple labs dating back to the 1960s. Now, a single Air Force laboratory consolidated all S&T programs under the roof of one streamlined organization. Transitioning to the new lab was not easy. It happened because people worked hard to make the implementation process a success—something General Paul knew better than anyone else: "I'm proud of the job they did. They really stepped up to the challenge." Clearly, one challenge had ended. But lab people had little time to dwell upon this watershed event. Air Force Research Laboratory leaders now had the new challenge of taking the lead in developing technology that would enable the most capable weapon systems to support US war fighters in the twenty-first century.[22]

Notes

1. Briefing, Dr. Brendan B. Godfrey, subject: D.C. Presence Task Group, 7 May 1997.

2. Gen Henry Viccellio Jr., interviewed by author, 24 June 1998; and Maj Gen Richard R. Paul, interviewed by author, 2 March 1998.

3. Maj Gen Richard R. Paul to AFMC/CC (General Viccellio), letter, subject: Single Laboratory Headquarters Location, 7 February 1997.

4. Briefing, Dr. Brendan B. Godfrey, subject: D.C. Presence Task Force Kickoff Meeting, 18 February 1997; and briefing, Dr. Brendan B. Godfrey, subject: Status Report: D.C. Presence Task Group, 22 March 1997.

5. Godfrey briefings, 18 February and 22 March 1997; briefing, Maj Gen Richard R. Paul, subject: Air Force Research Laboratory: Status Report, June 1997; and message, General Paul's web site, subject: AFRL Phase II: A Decision Flowchart; on-line, Internet, 25 April 1997, available from http://stbbs.wpafb.af.mil/STBBS/labs/single-lab/updates.htm.

6. Background paper, "Washington Office Director" [March 1997], in D.C. Presence Task Group Summary notebook.

7. Allan C. Schell to Maj Gen Richard R. Paul, letter, subject: Report of the Laboratory Alumni Review Team, 23 June 1997, with attached Laboratory

Alumni Review Team report; briefing, Dr. Bill Berry, subject: CONOPS [Concept of Operations]/Staffing for AFRL/D.C., 29–30 September 1997; message, General Paul's web site, subject: AFRL Phase II Status; on-line, Internet, 31 July 1997, available from http://stbbs.wpafb.af.mil/STBBS/labs/single-lab/updates.htm; and message, General Paul's web site, subject: Key Leaders for the AFRL Phase II Organization; on-line, Internet, 22 August 1997, available from http://stbbs.wpafb.af.mil/STBBS/labs/single-lab/updates.htm.

8. AFRL Organization Change Request Package, appendices A–L, August 1997.

9. Ibid.; message, General Paul's web site, subject: Midcourse Heading Check; on-line, Internet, 24 March 1997, available from http://stbbs.wpafb.af.mil/STBBS/labs/single-lab/updates.htm; and message, General Paul's web site, subject: AFRL Corporate Information Office; on-line, Internet, 7 May 1997, available from http://stbbs.wpafb.af.mil/STBBS/labs/single-lab/updates.htm.

10. AFRL Organization Change Request Package; and Maj Gen Richard R. Paul, notes, subject: Lab Reorganization, 20 April 2000.

11. AFRL Organization Change Request Package; and Paul notes.

12. AFRL Organization Change Request Package.

13. Ibid.

14. Ibid.; briefing, Maj Gen Richard R. Paul, subject: Air Force Research Laboratory: Status Report, August 1997; and Paul notes.

15. Message, General Paul's web site, subject: Getting Approval of AFRL Phase II; on-line, Internet, 13 August 1997, available from http://stbbs.wpafb.af.mil/STBBS/labs/single-lab/updates.htm; and briefing charts, subject: AFRL Phase II Organizational Change Request [August 1997], in AFRL notebook no. 29.

16. Message, General Paul's web site, subject: Personnel Allocation Process; on-line, Internet, 16 September 1997, available from http://stbbs.wpafb.af.mil/STBBS/labs/single-lab/updates.htm.

17. Message, General Paul's web site, subject: AFRL Status—What's the Deal?; on-line, Internet, 10 October 1997, available from http://stbbs.wpafb.af.mil/STBBS/labs/single-lab/updates.htm; and Paul notes.

18. Message, General Paul's web site, subject: Commander's Corner; on-line, Internet, 22 October 1997, available from http://stbbs.wpafb.af.mil/STBBS/labs/single-lab/updates.htm.

19. Ibid.

20. Headquarters Air Force Materiel Command, Special Orders GA-1 and GA-2, 29 October 1997.

21. Col Jacob Kessel, chief, Manpower and Organization, AFMC Directorate of Plans and Programs, to AFRL/XPM, letter, subject: Air Force Research Laboratory, Phase II, Reorganization, 30 October 1997.

22. Robert Ely, AFMC Public Affairs, "Air Force Consolidates Research under New Lab Organization," *Kirtland AFB Focus*, 31 October 1997.

Chapter 15

Conclusion

President Lyndon B. Johnson considered himself a pragmatist who liked to size up people quickly. According to his Texas yardstick of humanity, people fitted into one of two real-world categories. There were those, as he put it, who "do"—the hardy individuals of strong convictions who would fling themselves into the midst of controversy and take a fervent stance on the difficult problems of the day. Above all, they possessed a passion and determination to bring about meaningful change. In pursuit of that goal, they were not afraid to take risks. To them, reform was not some lofty, intellectual vision reserved for some more distant and appropriate time. Rather, they had a greater sense of urgency to get on with things now. Although the "doers" understood and did not discount the importance of vision, they placed much more emphasis and value on those individuals actively engaged in the hands-on work to transform the current inferior state of affairs into a more productive system.

Unfortunately, a large share of the population, in Johnson's estimation, did not fit into the ranks of the doers when it came to making radical changes in government. Others fell into the second category of "nondoers"—typically the planners, staffers, organizers, consultants, managers, visionaries, and facilitators who created grandiose schemes but depended almost exclusively on others, the doers, to get the job done right and on time. Johnson especially liked to jab the bright Harvard boys and East Coast intellectuals by educating them on the realities of politics that they would not necessarily glean from their classroom studies. He didn't hesitate to offer his observations on the ordering of the universe when he declared, "Those who can, do. Those who can't, write books about it. For it is far easier to pontificate about defense and foreign politics than it is to implement them."[1]

Raised in the hard-knocks school of the Texas hill country, Johnson measured people more by what they did as opposed to what they said they would do. He had little respect for people who retreated to the comfort of the sidelines and then became the most vocal critics of those who had to make the tough decisions. These sideline experts always seemed to have a more logical and better solution to every problem, but when it came to the moment of truth, they were unwilling to enter the arena and put their idea to the test by marshaling the forces required to implement their superior plan.

Despite Johnson's propensity to overuse hyperbole to make a point, much of what he believed in could be applied to the evolution of the Air Force Research Lab. The single laboratory represented a study in microcosm of a major government reform movement that produced radical change and upheaval in the military's S&T community. Consolidation of the Air Force's four laboratories into one could not have happened without a legion of doers. Unlike their position in Johnson's perspective on the world, the transition to a consolidated laboratory relied heavily on the talents of many military and civilian "planners." In truth, creation of the new lab required the combined expertise of planners and doers alike.

In this case, the line between planner and doer often was blurred. For example, 13 task groups focused on planning functions—studying a specific problem and developing various options for setting up the lab. Although study and analysis led to several possible "plans," group members also had to take action by selecting what they believed was the best plan. Thus, they became doers by proposing to General Paul and the Corporate Board the specific actions necessary to expedite the formation of the laboratory. Paul and the board members then repeated the process by studying and analyzing, selecting the best option, and then making the decision to move forward with that plan. Incorporating that action into the OCR and securing its approval gave General Paul the go-ahead to establish the lab.

Action rather than endless planning distinguished the transition effort. From the start, Paul and his senior staff made a conscious effort to instill an environment of "doing" rather than excessive planning and of encouraging people to move ahead with confidence. Importantly, the task groups' recommendations, the Corporate Board's conclusions, and General Paul's web site and road shows all showed very specific signs of progress. The transition operation succeeded because General Paul realized he could not afford to allow the process to bog down by studying issues to death. Thus, only four short months after establishing the task groups, he shifted the focus from planning to implementing the consolidated lab—another outward sign of progress.[2]

Certainly a significant accomplishment—and one too easily overlooked—is the speed with which this process occurred. After Secretary Widnall gave her approval on 20 November 1996, the new lab found itself up and operating in only 11 months—a monumental achievement, considering all the confusion, false starts, geographic separation of the existing labs, and a host of other complexities associated with building a new unit from scratch. Some of the credit for this must go to General Paul, who, from the beginning, set his priorities for standing up the lab in October. In the face of resistance to change and some heavy criticism, both inside and outside the laboratory system, he remained determined—especially after consulting with the director of the Army Research Laboratory—to adhere to his "90 percent rule" and use his leadership skills to keep the lab on schedule.

Clearly, General Paul did not have a monopoly on strong leadership traits. Leadership exhibited by other individuals—both military and civilian—at all levels throughout the lab also contributed to the project's success. Tech directors, division and branch chiefs, support staff, review board members, and numerous other dedicated workers—some of whom Paul did not even know—all took on and completed complicated taskings. As with any military operation of this magnitude, decisions did not satisfy everyone, some things did not go exactly as planned, and some people failed to see how their tasks fit into the big picture. More importantly, though, senior leaders made decisions and took action to overcome inertia rather than engaging in endless debate that would have delayed creation of the single lab.

In the midst of trying to bring about fundamental change to the S&T arm of the Air Force, well-meaning critics continued to try to introduce doubt into Paul's mind. The independent review boards wanted him to slow down, reevaluate, and conduct more study and analysis before moving forward. Individuals who either had strong ties to the old lab system or saw themselves as prominent players in the new lab asked question after question: Does this process simply rearrange the deck chairs to create 10 separate labs rather than one? Was AFRL a resurrection of the old Directorate of Laboratories, which existed under the former Air Force Systems Command? Did the new headquarters XP have too much power and add another level of unnecessary bureaucratic layering? Wasn't the notion of "doing more with less," although a catchy slogan, unrealistic? Would the new lab infrastructure severely diminish career progression for military personnel and civilians? The AFRL commander responded "no" to all these questions and proceeded with his implementation plan.

Paul's senior staff, four lab commanders, and tech directors provided substantial support to the AFRL commander during these difficult times. Although they did not agree with him on every issue, they never doubted his sincerity or motives for moving to a consolidated laboratory. Paul used his staff and commanders to develop a logical implementation plan, insisting that task groups representing a broad cross section of the organization study the options and make recommendations. He listened to what the task groups recommended, made himself available, and engaged in discussions with them to support or counter their ideas. Not only did he value the opinions of his Corporate Board and three independent review boards, but he also sought input from the employees who attended his road shows, encouraging everyone to comment through his web site. Clearly, General Paul did not operate in isolation.

But one thing he could not share. After listening to and weighing all sides of the issues, he alone would make the final decisions. By July he had done so, effectively ending further debate. At this point, the senior staff and four commanders closed ranks to support their commander—as would any disciplined military unit determined to conduct and complete its mission.

Creation of the single laboratory did in fact bring about fundamental changes—and with minimal disruption to personnel. Indeed, most employees retained their positions and locations, and no organizations physically moved. At the same time, everyone knew that personnel and budget cuts would occur over the next few years—after all, these actions played a part in creating the lab. There was just no getting around these two issues as the new lab moved on to phase IIB of the implementation plan.

In the final analysis, Air Force leaders accomplished what they had set out to do by consolidating four laboratories into one, thus creating a more streamlined infrastructure designed to save money and reduce the size of the workforce over the long haul. People and dollars now came under the control of a single lab commander who could more effectively and efficiently develop an investment strategy more responsive to customer needs. Moreover, the new lab organization made major strides in reducing the fragmentation of similar technologies previously distributed among multiple directorates. Above all, formation of the Air Force Research Laboratory in October 1997 was a bold and irreversible first step in positioning the Air Force on the cutting edge of science and technology. No one should doubt that this new organization will take the lead in developing enabling technologies to keep our Air Force superior and ensure the nation's defense as an unsettled world moves through the new millennium.

As Air Force Research Lab leaders peer into the future to deal with the enormous challenges and responsibilities ahead, it will be useful for them to turn their eyes to the past for guidance on how best to proceed through unmarked territory. The lesson is that people facing seemingly insurmountable odds make the difference in any great endeavor. Although AFRL leaders will face daunting and unknown obstacles in the years to come, they can gain a sense of quiet confidence from the inspirational words of Theodore Roosevelt, spoken a hundred years ago:

> Far better it is to dare mighty things, to win glorious triumphs, even though checkered by failure, than to take rank with those poor spirits who neither enjoy much nor suffer much, because they live in the gray twilight that knows not victory nor defeat.[3]

Notes

1. Kenneth L. Adelman and Norman R. Augustine, *The Defense Revolution: Strategy for the Brave New World* (San Francisco: Institute for Contemporary Studies Press, 1990), 197.

2. Message, General Paul's web site, subject: The AFRL Road Ahead; online, Internet, 24 July 1997, available from http://stbbs.wpafb.af.mil/STBBS/labs/single-lab/updates.htm.

3. Theodore Roosevelt, speech before the Hamilton Club, Chicago, 10 April 1899.

APPENDIX A

Chronology

1945	First meeting of the Scientific Advisory Group—forerunner of the Air Force Scientific Advisory Board (SAB).
1955	A-76 guidance first appears in Bureau of the Budget bulletins: The A-76 process determines whether contracting government work would be more cost-effective than employing government workers to perform the same jobs.
1966	Office of Management and Budget publishes Circular A-76.
1982	Air Force Systems Command assigns laboratories to centers that report to "product divisions."
1986	Packard Commission's blue-ribbon study looks at ways to operate the Department of Defense (DOD) in a more efficient and economical manner. David Packard, former undersecretary of defense, heads the study, which focuses on reforming four core areas: national security planning/budgeting, military organization/command, acquisition organization/procedures, and government-industry accountability.
June 1986	Packard Commission's blue-ribbon study issues its final report, *A Quest for Excellence*, proposing sweeping reforms to improve efficiency and save money in DOD.
1986	President Ronald Reagan signs National Security Decision Document 219, directing implementation of the major recommendations of the Packard Commission.
1986	President Reagan signs into law the Goldwater-Nichols Department of Defense Reorganization Act, considered the most significant defense-reform effort since 1947.

1986	Position of undersecretary of defense for acquisition created as result of the Goldwater-Nichols Act. The new undersecretary sets overall DOD procurement and research and development (R&D) policy and provides centralized control over all acquisition programs.
Late 1980s	The political/military pendulum begins to swing in the opposite direction. With the fall of the Berlin Wall in 1989 and the dissolution of the Soviet Union in 1991, unpredictable regional conflicts replace the monolithic communist threat.
July 1988–July 1992	Col Richard R. Paul serves as commander of Wright Lab, gaining invaluable, practical day-to-day experience directing major technology programs for advancing aerospace systems.
1989	President George Bush hears complaints from congressional representatives who feel that the services are dragging their feet in supporting management reforms initiated by the Packard Commission and the Goldwater-Nichols Act.
February 1989	President Bush directs Secretary of Defense Dick Cheney to draft a plan to look at ways to improve management (with fewer employees) and organizational efficiency in DOD. The goal is to devise a strategy to implement sweeping reforms proposed in the Packard Commission's report.
12 June 1989	Secretary of Defense Cheney completes a major reorganization plan known as the Defense Management Review (DMR) that addresses ways to improve the defense procurement process. It urges military services to borrow/implement streamlined business practices used in the private sector.

July–September 1989	As part of the continual defense management review process following the DMR report of June 1989, Secretary of Defense Cheney appoints special groups to investigate options for consolidating DOD functions, including laboratories.
30 October 1989	As a result of the special study groups, DMR Decision 922 strongly advises that the Pentagon consider merging all military laboratories directly under DOD.
October 1989–April 1990	DMR Decision 922 directs John A. Betti, undersecretary of defense for acquisition, to conduct an extensive study that focuses on the advantages/disadvantages of inter- and intraservice consolidation of laboratories. Betti asks Dr. George P. Millburn, deputy director of defense research and engineering in the Pentagon, to work with three services to explore the entire range of laboratory options.
30 April 1990	Dr. Millburn reports his findings to Betti and provides various possible solutions, including reducing the number of labs and combining all service labs into one DOD laboratory.
13 December 1990	Thirteen Air Force laboratories become four.
13 December 1990–31 October 1997	During this period, the organizational structure of each laboratory is, to a large degree, autonomous.
July 1992	Systems and Logistics commands merge to form Air Force Materiel Command.
July 1992	Gen Richard R. Paul is selected to serve as director of Science and Technology (S&T) at Air Force Materiel Command. He is responsible for leading and devising investment strategy

covering the full spectrum of Air Force technology activities.

October 1992	Consolidated Army Research Laboratory headquartered at Adelphi, Maryland, is formed.
FY 1993–FY 1996	During this time, 19 A-76 studies are completed in DOD.
1993	Dr. George R. Abrahamson, chief scientist of the Air Force, leads a study entitled "The Blue Ribbon Panel on Management Options for Air Force Laboratories."
1993	The Clinton administration appoints John Deutch and Adm David E. Jeremiah, vice chairman of the Joint Chiefs of Staff, to co-chair the Infrastructure Review Panel, which will examine the possibility of reducing DOD's infrastructure and its effect on the permanent workforce.
1993	DOD's Bottom-Up Review (BUR) study assesses options for reducing DOD's labor force as the nation shifts away from a strategy designed to meet the Soviet global threat to a new one based on preventing aggression by regional powers. Out of BUR comes a new strategy of engagement and enlargement that calls for US forces to work with regional allies to fight and win two major regional conflicts that occur nearly simultaneously.
23 November 1993	President William Jefferson Clinton announces a plan for an across-the-board review of all federal laboratories to streamline laboratory operations in view of projected decreases in federal R&D dollars.
17 December 1993	Gen Ronald W. Yates, commander of Air Force Materiel Command, verbally expresses

his reservations about the findings put forth by Dr. Abrahamson's blue-ribbon panel.

February 1994

Since the military services show no inclination to voluntarily reduce civilians, Secretary of Defense Les Aspin convenes a meeting with his deputy, William J. Perry, and Edwin Dorn to discuss alternatives.

3 February 1994

Secretary of Defense Aspin leaves his post.

5 May 1994

President Clinton issues a directive establishing the Interagency Federal Laboratory Review.

May 1994– May 1995

At the direction of President Clinton, the National Science and Technology Council (NSTC) reviews the nation's three largest laboratory systems operating within DOD, the Department of Energy, and the National Aeronautics and Space Administration.

2 June 1994

Edwin Dorn, undersecretary of defense for personnel and readiness, sends a letter to all secretaries of the military departments with year-by-year projections of how many civilian positions each service has to remove. This becomes known as the infamous "Dorn memo" directing each service to reduce 4 percent of its civilian workforce each year from 1994 through 1999. In 2000, a 3 percent civilian reduction will be imposed, followed by a 2 percent reduction in 2001. The goal is to reduce civilian positions by almost 30 percent by the end of 2001.

November 1994

Gen Ronald R. Fogleman, chief of staff of the Air Force, and Secretary of the Air Force Sheila E. Widnall send a letter to Dr. Gene McCall, chairman of the Air Force SAB to remind him of the enormously significant role Gen Hap Arnold and Dr. von Kármán played after World War II in establishing and promoting the importance of S&T in developing the Air Force of the future.

Mid-1990s   The "planning, studying, and assessing" phase of the laboratory structure ends. The Clinton administration, Congress, and DOD want the Air Force to take action and start reconfiguring the labs to produce an even leaner and more cost-effective R&D operation.

1995   DOD's Base Realignment and Closure (BRAC) exercise.

May 1995   Air Force Secretary Widnall and General Fogleman initiate a movement to produce a new Air Force vision for the twenty-first century. Gen Thomas S. Moorman Jr., Air Force vice chief of staff, chairs a study group consisting of senior military and civilian leaders to come up with a more realistic vision that would be responsive to changing political conditions around the world.

May 1995   Office of the Secretary of Defense releases its defense planning guidance. Budget officials use this information as building blocks to prepare the Program Objectives Memorandum for fiscal year 1997.

15 May 1995   After a year of investigating how the three laboratories operate, the NSTC submits its final report to President Clinton, indicating unanimous support for the legitimacy of laboratory functions but ample opportunity to improve management and cut redundancy.

June 1995   Gen Henry Viccellio Jr. assumes command of Air Force Materiel Command, Wright-Patterson Air Force Base, Dayton, Ohio.

December 1995	The Air Force publishes *New World Vistas* report to coincide with and commemorate the 50th anniversary of von Kármán's *Toward New Horizons*.
February 1996	The Air Force identifies 540 positions across the four labs as eligible to undergo the A-76 evaluation process.
10 February 1996	Congress passes the National Defense Authorization Act for fiscal year 1996. Section 277 of this legislation (Public Law 104-106) directs the secretary of defense to develop a five-year plan to consolidate and restructure laboratories and test and evaluation centers assigned to DOD.
15 February 1996	President Clinton instructs the secretary of defense to submit a report to him by this deadline, "detailing plans and schedules for downsizing the DOD laboratories."
March 1996	Office of Management and Budget publishes a new version of Circular A-76's *Revised Supplemental Handbook*.
Spring 1996	Col Dennis Markisello, military deputy for the Science and Technology Directorate, joins Don Daniel, Tim Dues, and Vince Russo as the fourth key figure brought into General Paul's inner planning cell to assist him in developing a new lab infrastructure.
30 April 1996	The secretary of defense delivers *Vision 21* plan to Congress, which becomes the catalyst for the Air Force to move forward to completely revamp the laboratory system. The final *Vision 21* report combines outcomes from two studies—the DOD report compiled as a result of NSTC recommendations on laboratory restructuring and the *Vision 21* DOD

plan prepared in response to the Defense Authorization Act of 1996.

Late Spring 1996   The genesis of General Paul's idea of proposing a single lab takes place on an airplane during a return trip with General Viccellio. General Paul provides a verbal explanation of the numerous advantages of going to a single lab. General Viccellio likes the idea and asks General Paul to give him more specifics about what the new laboratory structure would look like.

Summer 1996   Joint Chiefs of Staff publish *Joint Vision 2010.*

Fall 1996   General Moorman's vision team issues its new strategic blueprint called *Global Engagement,* comprised of six core competencies: air and space superiority, rapid global mobility, precision engagement, information superiority, global attack, and agile combat support.

8–12 October 1996   Lt Gen Lawrence P. Farrell Jr. presents a number of laboratory organizational options to the Air Force's top leaders gathered at a five-day Corona conference held at the Air Force Academy in 1996. Secretary Widnall and General Fogleman agree that the single-lab option is the way to go.

20 November 1996   Blaise Durante briefs Secretary Widnall on the single-laboratory proposal, the Air Force's answer to *Vision 21.* The secretary approves the single lab.

22 November 1996   General Paul sends a letter to four lab commanders, the AFOSR director, and his staff to officially announce that the secretary of the Air Force has approved the creation of a single laboratory.

Late November 1996	General Paul and Dr. Russo meet with Dr. John W. Lyons, director of the Army Research Laboratory, to find out first-hand how Lyons and his staff set up their consolidated lab. General Paul leaves the meeting knowing that he definitely does not want his transition team investing the majority of its time solving "the last" 10 percent of the Air Force lab's consolidation problems. Consequently, Paul decides to pursue the "90 percent solution," knowing ahead of time that mistakes will occur along the way.
5–6 December 1996	General Paul holds an off-site meeting at the Bergamo Conference Center near Wright-Patterson AFB.
Mid-December 1996	General Viccellio asks General Paul to provide a "heading check" on progress in forming the single lab.
31 December 1996	General Paul presents the overall long-range plan ("heading-check briefing") for establishing the single lab to General Viccellio, who fully endorses the phased approach for implementing the single lab.
January 1997	General Moorman explains the new vision to the Air Force Association: "The context of the long-range plan is built around sustaining our *core competencies* and reinforcing the central themes found in the strategic vision."
6 January 1997	General Paul forms transition organization. General Viccellio officially announces approval of the phased approach to single-lab implementation and directs Headquarters AFMC Command/Science and Technology Directorate to proceed.

13 January 1997	General Paul selects Dr. Russo to serve as the single-lab transition director and approves the lab-transition structure.
Early February 1997	Dr. Robert R. "Bart" Barthelemy, Technology Directorate Task Group, identifies "The Twelve" people responsible for drafting the first cut at determining how to form the new tech directorates.
3 February 1997	Maj Gen Michael C. Kostelnik, director of Plans at AFMC, submits the OCR package to Headquarters Air Force with a request to approve it by March 15.
Mid-February 1997	By this time, all members of the task groups have been named.
25–26 February 1997	Dr. Barthelemy calls the first meeting of the technology group at the Hope Hotel near Headquarters AFMC at Wright-Patterson AFB.
End February– Early March 1997	Task group kick-off meetings begin.
11 March 1997	General Paul's web site announces the "Getting Under Way" process.
18 March 1997	General Paul's web site announces the "Single-Lab Transition Organization."
19 March 1997	General Paul's web site announces "Single-Laboratory Road Shows," along with his solicitation of questions prior to his visits to lab locations.
20 March 1997	General Paul's web site announces "Corporate Board Meeting" for 20–22 March.
20–22 March 1997	Midcourse review of task forces/task groups briefs General Paul and the Corporate Board.

21 March 1997	Dr. Barthelemy provides the first tech-task-group heading check to General Paul.
23 March 1997	General Paul's web site announces "Single-Lab Task Groups and Focus Groups."
24 March 1997	General Paul selects the design for the new AFRL emblem.
24 March 1997	General Paul's web site announces "Mid-course Heading Check" and the decision to have a corporate information officer.
End of March–Early April 1997	Independent and Grassroots review boards provide input to the Corporate Board.
End of March 1997	Completion of phase I, a three-month operation calling for the stand-up of the interim laboratory organization.
1 April 1997	General Paul's web site announces the draft design of AFRL heraldry (emblem for the new organization).
1 April 1997	Special Order GA-9 announces the activation of Headquarters AFRL and assigns it to AFMC. The order also relieves the four laboratories and AFOSR from their assignment to their present organizations and reassigns them to AFRL, effective 8 April 1997.
1–2 April 1997	Grassroots Review Team meets with General Paul and his staff.
4 April 1997	General Paul's web site addresses "Existing Four Laboratories' Name and Heritage" and responds to concerns people expressed during the road shows about losing the labs' heritage due to organizational consolidation. Confirms that history and lineage of Air Force labs will be memorialized.

8 April 1997	By order of the secretary of the Air Force, the Air Force Research Laboratory is activated, signifying the end of phase I (interim organization). A "short and modest" ceremony is held at Wright-Patterson honoring this event. General Viccellio reads the orders activating the lab and unveils the new AFRL emblem. General Paul briefs the AFMC commanders.
8 April 1997	Independent Assessment Board convenes.
9–10 April 1997	Task groups for Support as well as Integration and Operations meet.
14 April 1997	General Paul's web site addresses the "Grassroots Review Panel," declaring it an "unqualified success!"
14 April 1997	General Paul's web site addresses the "Air Force Research Laboratory Independent Assessment Board Review."
17 April 1997	General Paul's web site addresses the completion of "Road Show I," beginning at Brooks AFB and completing a nine-site briefing circuit including Wright-Patterson, Hanscom, Rome, Kirtland, Edwards, Bolling, Eglin, and Tyndall Air Force Bases.
23 April 1997	General Paul sends a letter to General Viccellio requesting his approval for the new AFRL emblem in preparation for the phase II stand-up.
23 April 1997	General Paul forms the product-executives Red Team.
24 April 1997	Dr. Barthelemy makes second presentation to the Corporate Board.
25 April 1997	General Paul's web site addresses "AFRL Phase II: A Decision Process Flowchart."

Within the flowchart, each decision point is identified with a Corporate Board and a decision output, which feed into the next critical decision point.

7 May 1997	General Paul's web-site memo addresses "The AFRL Corporate Information Office."
7 May 1997	Corporate Board reviews the options of the product executives and Washington presence.
8 May 1997	General Paul briefs concepts (i.e., Technology Directorates, Plans and Programs, Operations and Support, Product Executives, etc.) to the current 22 technology directors.
12–13 May 1997	General Paul, Vince Russo, and other senior staff members attend an off-site meeting to start building a single-lab strategic plan.
14 May 1997	General Paul's web-site memo provides an "AFRL Progress Report—Directorates and More," announcing the framework for the technology directorates.
27 May 1997	General Paul's web site addresses "Organizational Development in AFRL."
End of May 1997	First version of the strategic plan lays out the vision and mission for the new lab.
2d week of June 1997	General Paul posts a message on the web site informing people of the progress of the selection of the new lab leaders. He explains that he is in the "final phases of proposing a set of provisional leaders."
12 June 1997	General Paul's web site provides a "Status Report" of phase II.
19 June 1997	In compliance with AFR 38-101, General Paul sends an OCR proposing the creation

of a single Air Force lab to the director of plans at AFMC.

24 July 1997	General Paul's web site addresses "The AFRL Road Ahead," with a preview of coming attractions about the reorganization.
31 July 1997	General Paul's web site announces "AFRL Phase II Status," including organizational structure and consolidation of 22 technology directorates to nine.
4 August 1997	General Paul's web site addresses the "Transition from AFRL Phase I to Phase II," including the impact on each laboratory.
7 August 1997	General Paul's web site addresses the "Office Symbols for AFRL Phase II," including new office symbols for the technology directorates.
7 August 1997	Gen George T. Babbitt, AFMC commander, signs the OCR and submits it to AF/XP later that day.
13 August 1997	General Paul's web site addresses "Getting Approval of AFRL Phase II."
22 August 1997	General Paul announces "Key Leaders for the AFRL Phase II Organization."
16 September 1997	General Paul's web site addresses "The Personnel Allocation Process," including his personal assurances that "my original intention still remains to minimize personal disruption and geographical moves to the extent possible."
October 1997	Completion of phase IIA, a six-month process representing the final organization at full manning, results in the assigning of personnel positions to one lab unit-manning document.

10 October 1997	General Paul's web site addresses "AFRL Status—What's the Deal?" The staffing process at the Pentagon takes longer than anticipated, and the OCR awaits the vice chief of staff's approval. General Paul believes that the OCR will be signed by the end of October.
22 October 1997	"Commander's Corner" on General Paul's web site announces, "It's official! The Secretary of the AF approved the AFRL reorganization today!" Announces that 31 October 1997 will be the "official stand-up date."
31 October 1997	The Air Force activates the Air Force Research Laboratory, marking the completion of phase IIA.
October 1997–1 January 2001	Phase IIB is designed to complete the "end-state" lab—the final organization at reduced manning.
1 April 1998	Secretary of Defense William J. Perry sets this date as an initial milestone for a five-year plan on how DOD laboratories would restructure.
May–June 1998	Secretary Perry and staff review the five-year plan on how DOD laboratories would restructure, in preparation for submitting the plan to President Clinton.
1 July 1998	Secretary Perry establishes target date to submit five-year plan to President Clinton on how DOD laboratories would restructure.
October 2000	Target date established by Secretary Perry to begin implementation of the laboratory five-year restructuring plan.
1 October 2005	Target date established by Secretary Perry for completion of the five-year plan.

APPENDIX B

Laboratory Consolidations
Prior to Formation of a Single Laboratory

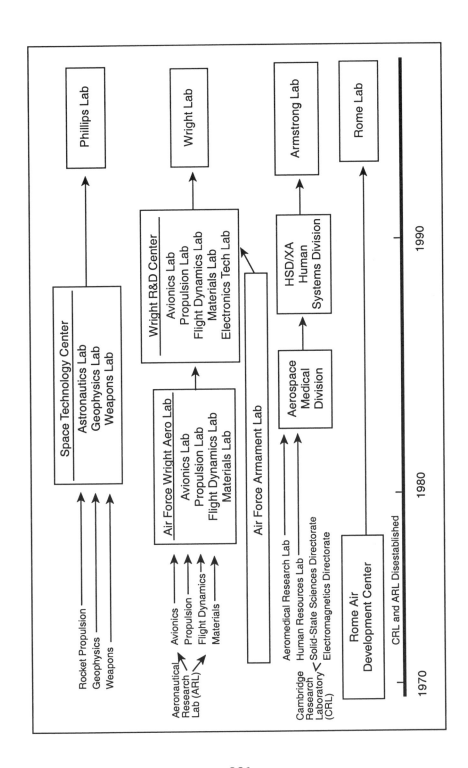

APPENDIX C

**Orders Activating the
Air Force Research Laboratory**

DEPARTMENT OF THE AIR FORCE
HEADQUARTERS AIR FORCE MATERIEL COMMAND
WRIGHT-PATTERSON AIR FORCE BASE, OHIO

Special Order   1 April 1997
GA-9

1. HQ Air Force Research Laboratory (HQ AFRL) is activated at Wright-Patterson AFB, Ohio, and assigned to HQ Air Force Materiel Command, effective 8 April 1997. Authority: DAF/XPM letter 918r, 31 March 1997, Organization Actions Affecting Certain Air Force Materiel Command Units, and AFI 38-101.

Mailing Address

HQ Air Force Research Laboratory (HQ AFRL)
4375 Chidlaw Road, Suite 6
Wright-Patterson AFB OH 45433-5006

2. Air Force Office of Scientific Research (AFOSR), Bolling AFB, D.C., is relieved from its present assignment to HQ Air Force Materiel Command, and is assigned to Air Force Research Laboratory, effective 8 April 1997. Authority: AFI 38-101.

3. Armstrong Laboratory, Brooks AFB, Texas, is relieved from its present assignment to Human Systems Center, and is assigned to Air Force Research Laboratory, effective 8 April 1997. Authority: AFI 38-101.

4. Phillips Laboratory, Kirtland AFB, New Mexico, is relieved from its present assignment to 377 Air Base Wing, and is assigned to Air Force Research Laboratory, effective 8 April 1997. Authority: AFI 38-101.

5. Rome Laboratory, Rome, New York, is relieved from its present assignment to Electronic Systems Center, and is assigned to Air Force Research Laboratory, effective 8 April 1997. Authority: AFI 38-101.

6. Wright Laboratory, Wright-Patterson AFB, Ohio, is relieved from its present assignment to Aeronautical Systems Center, and is assigned to Air Force Research Laboratory, effective 8 April 1997. Authority: AFI 38-101.

FOR THE COMMANDER

Jacob Kessel

JACOB KESSEL, Colonel, USAF
Chief, Manpower and Organization
Directorate of Plans

Distribution
1 - HQ USAF/SG, Wash DC 20330-5133
1 - HQ USAF/DPG, Wash DC 20330-1040
1 - HQ USAF/JAEC, Wash DC 20330-5120

SO GA-9

2 - HQ USAF/XPMO, Wash DC 20330-1070
1 - HQ USAF/ILXB, Wash DC 20330-1480
1 - AFPCA/DOVR, Wash DC 20330-1600
1 - AUL/LDEA, Maxwell AFB AL 36112-5564
1 - AFHRA/RS, Maxwell AFB AL 36112-6424
1 - HQ AFMC/HO/JA/PA/SCDP/ST/XPMO/
  XPMQ/XPMR
1 - AFMC QMIO
1 - HQ AFMC/DP/DPA/DPC/DPO
1 - AFOSR/CC
1 - HSC/CC/DP/HO/JA/MQ/PA
1 - AL/CC
1 - SMC/CC/DP/HO/JA/MQ/PA
1 - PL/CC
1 - ESC/CC/DP/HO/JA/MQ/PA
1 - RL/CC
1 - ASC/CC/DP/HO/JA/MQ/PA
1 - WL/CC

SO GA-9

DEPARTMENT OF THE AIR FORCE
HEADQUARTERS AIR FORCE MATERIEL COMMAND
WRIGHT-PATTERSON AIR FORCE BASE, OHIO

Special Order   8 April 1997
GA-12

Special Order GA-9, this headquarters, 1 April 1997, paragraph 4, which reads *Phillips Laboratory, Kirtland AFB, New Mexico, is relieved from its present assignment to 377 Air Base Wing,* is amended to read *Phillips Laboratory, Kirtland AFB, New Mexico, is relieved from its present assignment to* ***Space and Missile Systems Center***. Authority: AFI 38-101.

FOR THE COMMANDER

Jacob Kessel

JACOB KESSEL, Colonel, USAF   Distribution
Chief, Manpower and Organization   1 - HQ USAF/SG, Wash DC 20330-5133
Directorate of Plans   1 - HQ USAF/DPG, Wash DC 20330-1040
                                         1 - HQ USAF/JAEC, Wash DC 20330-5120
                                         2 - HQ USAF/XPMO, Wash DC 20330-1070
                                         1 - HQ USAF/ILXB, Wash DC 20330-1480
                                         1 - AFPCA/DOVR, Wash DC 20330-1600
                                         1 - AUL/LDEA, Maxwell AFB AL 36112-5564
                                         1 - AFHRA/RS, Maxwell AFB AL 36112-6424
                                         1 - HQ AFMC/HO/JA/PA/SCDP/ST/XPMO/
                                               XPMQ/XPMR
                                         1 - AFMC QMIO
                                         1 - HQ AFMC/DP/DPA/DPC/DPO
                                         1 - AFOSR/CC
                                         1 - HSC/CC/DP/HO/JA/MQ/PA
                                         1 - AL/CC
                                         1 - SMC/CC/DP/HO/JA/MQ/PA
                                         1 - PL/CC
                                         1 - ESC/CC/DP/HO/JA/MQ/PA
                                         1 - RL/CC
                                         1 - ASC/CC/DP/HO/JA/MQ/PA
                                         1 - WL/CC

SO GA-12

DEPARTMENT OF THE AIR FORCE
HEADQUARTERS AIR FORCE MATERIEL COMMAND
WRIGHT-PATTERSON AIR FORCE BASE OHIO

U.S. AIR FORCE

1947 - 1997

Special Order
GA-1

29 October 1997

1. The following Research Site Detachments are activated at the locations indicated and assigned to the Air Force Research Laboratory, effective 31 October 1997. Authority: AFI 38-101.

Detachment	Location
Detachment 1	Wright-Patterson AFB, Ohio
Detachment 2	Tyndall AFB, Florida
Detachment 3	Hanscom AFB, Massachusetts
Detachment 4	Rome, New York
Detachment 5	Brooks AFB, Texas
Detachment 6	Eglin AFB, Florida
Detachment 7	Edwards AFB, California
Detachment 8	Kirtland AFB, New Mexico

2. The following Technology Directorate Detachments are activated at the locations indicated and assigned to the Air Force Research Laboratory, effective 31 October 1997. Authority: AFI 38-101.

Detachment	Location
Detachment 9 (Space Vehicles)	Kirtland AFB, New Mexico
Detachment 10 (Information)	Rome, New York
Detachment 11 (Munitions)	Eglin AFB, Florida
Detachment 12 (Directed Energy)	Kirtland AFB, New Mexico

FOR THE COMMANDER

Jacob Kessel

JACOB KESSEL, Colonel, USAF
Chief, Manpower and Organization
Directorate of Plans and Programs

Distribution
1 - HQ USAF/SG, Wash DC 20330-5133
1 - HQ USAF/DPG, Wash DC 20330-1040
1 - HQ USAF/JAEC, Wash DC 20330-5120
2 - HQ USAF/XPMO, Wash DC 20330-1070
1 - HQ USAF/ILXB, Wash DC 20330-1480
1 - AFPCA/DOVR, Wash DC 20330-1600
1 - AUL/LDEA, Maxwell AFB AL 36112-5564
1 - AFHRA/RS, Maxwell AFB AL 36112-6424
1 - HQ AFMC/HO/JA/PA/SCDP/XPM/XPMQ/XPMR
1 - AFMC QMIO
1 - HQ AFMC/DP/DPA/DPC/DPO
1 - ASC/CC
1 - ESC/CC
1 - HSC/CC
1 - SMC/CC
1 - 377 ABW/CC
1 - AFRL/CC
1 - Dets 1-12, AFRL

SO GA-1

Golden Legacy, Boundless Future... Your Nation's Air Force

DEPARTMENT OF THE AIR FORCE
HEADQUARTERS AIR FORCE MATERIEL COMMAND
WRIGHT-PATTERSON AIR FORCE BASE OHIO

U.S. AIR FORCE

1947 - 1997

Special Order   29 October 1997
GA-2

1. The following units are inactivated effective 31 October 1997. Concurrently, unit designation will revert to the Department of the Air Force. The Director of Personnel, HQ AFMC, will reassign personnel. Upon inactivation, consult AFI 84-101 to dispose of flags and other historic artifacts. Dispose of supplies and equipment per current directives. Dispose of organizational records and submit a final report under current directives. Authority: DAF/XPM letter 974r, 28 October 1997, Inactivation of Certain Air Force Materiel Command Units, and AFI 38-101.

Unit	Location
Armstrong Laboratory	Brooks AFB TX
Phillips Laboratory	Kirtland AFB NM
Rome Laboratory	Rome NY
Wright Laboratory	Wright-Patterson AFB OH

2. The Air Force Materiel Command Technology Transition Office (AFMC TTO) is inactivated at Wright-Patterson AFB, Ohio, effective 31 October 1997. Concurrently, unit designation will revert to the Department of the Air Force. The Director of Personnel, HQ AFMC, will reassign personnel. Upon inactivation, consult AFI 84-101 to dispose of flags and other historic artifacts. Dispose of supplies and equipment per current directives. Dispose of organizational records and submit a final report under current directives. Authority: DAF/XPM letter 974r, 28 October 1997, Inactivation of Certain Air Force Materiel Command Units, and AFI 38-101.

FOR THE COMMANDER

Jacob Kessel

JACOB KESSEL, Colonel, USAF
Chief, Manpower and Organization
Directorate of Plans and Programs

Distribution
1 - HQ USAF/SG, Wash DC 20330-5133
1 - HQ USAF/DPG, Wash DC 20330-1040
1 - HQ USAF/JAEC, Wash DC 20330-5120
2 - HQ USAF/XPMO, Wash DC 20330-1070
1 - HQ USAF/ILXB, Wash DC 20330-1480
1 - AFPCA/DOVR, Wash DC 20330-1600
1 - AUL/LDEA, Maxwell AFB AL 36112-5564
1 - AFHRA/RS, Maxwell AFB AL 36112-6424
1 - HQ AFMC/HO/JA/PA/SCDP/XPM/XPMQ/XPMR
1 - AFMC QMIO
1 - HQ AFMC/DP/DPA/DPC/DPO
1 - AFRL/CC
1 - ASC/CC/HO/MQ
1 - ESC/CC/HO/MQ
1 - HSC/CC/HO/MQ
1 - SMC/CC/HO/MQ
1 - 377 ABW/CC
1 - Dets 1-12, AFRL

SO GA-2

Golden Legacy, Boundless Future... Your Nation's Air Force

Glossary

AFB	Air Force base
AFI	Air Force instruction
AFMC	Air Force Materiel Command
AFMC/ST	Air Force Materiel Command, Science and Technology Directorate
AFOSR	Air Force Office of Scientific Research
AFRL	Air Force Research Laboratory
AFSC	Air Force Systems Command
ASD	Aeronautical Systems Division
BRAC	Base Realignment and Closure [Commission]
BUR	Bottom-Up Review
C^3I	command, control, communications, and intelligence
C^4I	command, control, communications, computers, and intelligence
CIO	Corporate Information Office
C^4ISR	command, control, communications, computers, intelligence, surveillance, and reconnaissance
CSAF	chief of staff of the Air Force
DDR&E	Director, Defense Research and Engineering
DMR	Defense Management Review
DMRD	Defense Management Report Decision
DOD	Department of Defense
DPG	defense planning guidance
DS	Operational Support
ERB	Executive Resources Board
FY	fiscal year
HSC	Human Systems Center
IAB	Independent Assessment Board

NSDD	National Security Decision Directive
NSTC	National Science and Technology Council
OCR	organization change request
OMB	Office of Management and Budget
OSD	Office of the Secretary of Defense
PEO	product executive officer
R&D	research and development
RDT&E	research, development, test, and evaluation
RIF	reduction in force
SAB	Scientific Advisory Board
SAE	service acquisition executive
SAF/AQ/AQR	Secretary of the Air Force for Acquisition
S&E	scientist and engineer
SES	Senior Executive Service
S&T	science and technology
TBD	to be determined
T&E	test and evaluation
TEO	technology executive officer
TTO	Technical Transition Office
XP	Plans and Programs

Index

305